digital soul

contents

preface

Most people have a short list of books that they discovered at pivotal points in their life and that changed the way they think in crucial ways. In a sense, this book began in the 1960s, when I picked up a copy of Dean Wooldridge's book *The Machinery of the Brain*.[1] In this and his two succeeding books, *The Machinery of Life* and *Mechanical Man: The Physical Basis of Intelligent Life*, Wooldridge laid out an astounding idea: that human beings function entirely according to the laws of physics.[2] This idea was both amazing and disturbing to me because it upset a lifetime of Catholic doctrine about an immortal soul and the spiritual nature of the human mind. In place of mysticism, it offered the possibility that the whole of human experience is not only understandable but reproducible. Could it possibly be true? I had to find out more.

My curiosity led me to books such as Huston Smith's *The Religions of Man*, B. F. Skinner's *Beyond Freedom and Dignity*, Edward O. Wilson's *On Human Nature*, Pamela McCorduck's *Machines Who Think*, Douglas Hofstadter's *Gödel, Escher, Bach*, and Robert Wright's *Moral Animal* as well as the 1988 Public Broadcasting Service series *Joseph Campbell and the Power of Myth*.[3]

I found out that the idea of man as a mechanism is an ancient one, but that this idea has been eclipsed by mystical and spiritual views of man promoted by the world's religions. I learned that science is discovering that the most complex human behaviors, including ethical and moral reasoning, are rooted in basic biological imperatives and that brain research is revealing more and more mental functions, like emotions and consciousness, as the workings of a wonderfully complex information processor.

Adding to the evidence for physical explanations of intelligent behavior are the great strides that have been made since the 1950s toward developing machines with *artificial intelligence,* or *AI.* As a result, machines that can learn are taking over more and more functions that we have always assumed to be uniquely human, such as playing championship chess. Given the present rate of growth of computing power, some scientists seriously predict that machines will become *more intelligent and aware than we are* in just thirty to forty years—outperforming humans in every important way, and perhaps even forming a global brain!

What are we to make of all this? Although we are still far from creating an artificial human being, it might be wise to start thinking about how the future intelligent machines could change the way we see ourselves and even alter humankind's future. A world populated with intelligent artifacts that have minds and feelings of their own raises many tricky questions, such as these:

- How would humans react on discovering that the club of sentient beings is not as exclusive as they thought?
- How would that knowledge change our moral and ethical values?
- How would it affect our notions of freedom and dignity?
- How would it affect our beliefs in God?
- Can computers be conscious?
- Can they have emotions?
- If so, what are their rights and responsibilities?
- If we make something that is indistinguishable from a person, should we treat it like one?

- Should we be worried that superintelligent machines will somehow take over the world?
- Are intelligent machines the next step in evolution?
- Will humans and machines somehow merge?
- Will humans become extinct? Or immortal?
- And most important . . . How much control do we have over the process?

This book is not for experts in computers or artificial intelligence. My audience is ordinary people who are curious enough to ask questions like these, and who want to be able to make informed decisions about the course of AI's development. It is even for those who think that all this is rubbish—that no matter how "smart" machines get, they will never truly think, have a soul, or be self-aware in the same sense we are. Whatever your leanings, it should do no harm to explore, poke around, ask questions, and try to find out what makes machines so smart, what their inherent limitations might be, and where the boundary between human and artificial intelligence might lie. If these questions make you vaguely uncomfortable, it may be because they challenge the very foundations of all our social, legal, and religious institutions.

Because there are already lots of books about the mechanics of AI, as well as works of fiction and nonfiction that speculate about future worlds inhabited by thinking machines, it is reasonable to ask what new territory I hope to cover here. We will explore an unfamiliar land along the boundary between science and human values. There, we will seek out the logical structure of human feelings, consciousness, and morality that would make it possible for machines to possess them as well. Then we will delve into the *social, moral, ethical, and religious consequences* of creating thinking and feeling artifacts.

What sort of consequences? As any science-fiction fan knows, the big problem with intelligent machines, going back as far as *Franken-*

stein, is loss of control. The public probably had its first taste of the moral and ethical aspects of computer behavior in Stanley Kubrick's 1968 movie *2001: A Space Odyssey.* The HAL 9000 computer in charge of running the *Discovery* spaceship suddenly turned on its crew, murdering all but one. This sort of "psychotic" behavior should come as no surprise when machines designed to be our servants become so complex that we can no longer understand how they work. If we put them in charge of critical aspects of our lives, then what is to stop them from taking over completely and pursuing goals that we can barely comprehend? And if they did so, how would superintelligent machines ultimately regard lowly humans? Would they even stoop to communicate with us?

How could we avoid such a grim future? Only by knowing what our choices are and by carefully thinking about the kind of future we want to live in can we hope to influence the science and technology policies that will take us there.

I wish to thank my friends and colleagues who let me pester them with my strange questions and who shared their insights with me. I also thank those who read the manuscript and made valuable suggestions, particularly my wife, Julianne Cassady; Bob and Janet Evans; Jay Palmer; Patricia Boyd; and Marilynn Breithart.

1

Artificial Intelligence—
That's the Fake Kind, Right?

Pay no attention to that man behind the curtain!
THE WIZARD OF OZ

Just about everyone has an opinion about the prospects of creating artificial intelligence and artificial life. These opinions range from *Are you crazy?* to *Why not?* Many regard the idea as blasphemous and accuse scientists of playing God. Others confine it to the realm of science fiction—entertaining but not to be taken seriously. Still others think it *can* be done but question whether it *should* be done. Most people believe that there is something intangible or spiritual about the way that minds work that can never be captured in silicon circuitry. The very term *artificial intelligence* suggests that machine intelligence will always fall short of *real* intelligence. Whatever your beliefs, you probably hold them with religious conviction. The mere mention of machines that are conscious, have feelings, or could have rights usually generates such heat and emotion as to preclude rational debate.

Why does the idea of AI stir up such emotion? Some people get nervous about AI when their egos won't allow them to recognize any kind of intelligence except the human kind. Their idea of human dignity depends on a natural superiority over all other creatures (and even over some other humans). Recent movies like *AI* and *Bicentennial Man* reinforce the idea that, although we might someday create machines that act human in many respects, they will always lack (and secretly long for) that intangible quality that would make them truly human.

Others might be called *carbon chauvinists*. They scoff at AI because they believe that only living things made from flesh and blood can exhibit intelligence. When they look at the "smart" machines of today, they seize upon their lapses and say, "See, I told you that a mere machine can never be as smart as a living creature."

Such emotions are hard to set aside, especially since they form the basis for many of our moral and ethical beliefs. Yet they also form a barrier to rational scientific inquiry. To investigate objectively whether AI is really possible, we must put aside the insecurities that lie at the root of such feelings and keep an open and curious mind. Does this mean that we have to check our dignity at the door if we want to ask questions like those posed in the preface? If we derive our dignity from a natural superiority over others, then the answer is probably yes. But this is the kind of dignity that we lost when Copernicus displaced Earth from the center of the universe, and when Darwin suggested that humans evolved from lower life-forms. Intellectual honesty is incompatible with a need to be at the center of the universe. It also requires us to ask questions without fear of where the answers may lead us.

The questions we want to ask challenge religious traditions, as well as New Age thinking. Both assert that the human mind—soul, spirit, what have you—lies beyond the limits of scientific examination. If you ask *why* it can't be studied scientifically, they say *It just can't! There are just some things that you can't analyze!* and *Subjective experience lies beyond the reach of scientific inquiry!* Besides begging the question, such answers are all saying essentially the same thing: There are certain questions that *shouldn't* be asked. Whenever someone doesn't want you poking into something, it's generally because the answers would threaten cherished beliefs or entrenched power structures. Which is fine. That's an honest reason. But there's a big difference between saying something *can't* be studied and something *shouldn't* be studied.

We will proceed in the belief that ignorance doesn't solve any problem, that mysticism explains nothing, and that there is nothing that can't be studied. We won't find all the answers, but this will not stop us from raising the questions.

Catching Up

The emergence of electronic computers since the 1950s has prompted an explosion of books and articles about "thinking machines" and their relation (if any) to the workings of the human mind.[1] But speculation and debate about the mechanical nature of the human mind are not new. They've been raging since the days of Aristotle and Plato. Automata and thinking machines of all sorts permeate human history.[2] The Egyptians had mechanical statues of gods that spoke, gestured, and prophesied. Today, no respectable science-fiction epic is complete without robots like R2D2 and C3PO of *Star Wars* and superintelligent computers, like HAL in *2001: A Space Odyssey*. Machines that think and act like humans are firmly established in our mythology.

How long will it take our science and technology to catch up with those fantasies? It may happen gradually or suddenly; in twenty years or two hundred years. It may be happening already. Intelligent machines may assume a form more or less as we imagine them now, or something entirely unexpected. In any case, we have already set off down that road, and there seem to be no exits. We are headed inexorably toward a future with intelligent machines doing more and more things that we now think of as uniquely human. Is it just a matter of time before there are machines that outperform humans in every important way?

Technology always perturbs our values and moral codes—the set of rules that guide the way we interact with each other. Fire, the wheel, the printing press, atomic energy, and genetic engineering—each made some of those rules obsolete. It is just as foolish to think that our moral and ethical codes are written in stone, as it is to imagine that knowledge of any of these technologies could somehow be suppressed or ignored. Citing our abuses of atomic energy, some say that we shouldn't even be thinking about intelligent machines, because we aren't morally and ethically ready to deal with the consequences of such an inquiry. And they would be right. We never are. But this is no reason to suppress scientific knowledge. Besides, even if we wanted to stop the development of intelligent artifacts, we couldn't. It's one of those ideas "whose time has come." But does this mean that we are powerless? That we have to passively watch as we become slaves to our technology? Of course not. It just means that once again we will be forced to rethink and overhaul outdated values and moral codes. This time, however, the changes will be so massive, they will upset the very way we think about ourselves.

Flawed Tools

We are handicapped in our inquiry by the tools available for describing our subject. Our very language of subjective experience, which is built on the notion of having a self, is full of loaded words that constrain and muddy our thinking. The pronouns *I* and *you* create images of autonomous agents. Linguistic traditions force us to think of *body* and *mind* as separate and distinct entities. Everyday notions like *free will* and *moral responsibility* contain underlying contradictions. Language also uses definitions and forms of the verb *to be* in ways that force us to think of classes of things as clearly defined (Is a fetus a human being or not?), when in fact every classification scheme has fuzzy boundaries and continuous gradations. In this book, I will argue that the distinction between "artificial" and "real" intelligence is merely a linguistic trap to be avoided.

One widely used tool for understanding intelligence is called *introspection*. At first, it seems to offer insights into mental processes by examining how they "feel" to us. This tool is flawed because the observing instrument itself is a part of what we are trying to explain. Such an instrument would most likely fool itself. What we are "conscious" or "aware" of at any given moment tells us nothing at all about how consciousness *works*. The feeling of consciousness, though "subjectively obvious," defies detection and analysis by conventional "objective" means, except that we know that it is strongly correlated with electrical activity in a certain part of the brain stem. The same can be said for feelings that we subjectively experience as *anger, love, happiness, grief, intuition,* or *creativity*. These labels could well be comfortable metaphors for complex genetic programs that create illusions, like the Wizard of Oz, for their own procreative purposes.

Another tool, used by brain researchers, maps the parts of the brain that are electrically or metabolically active when we feel certain emotions. Such maps, in their present state, are useful in diagnosing brain damage, but they are not the same as explaining how emotions and consciousness work, any more than an array of flashing lights on a computer panel tells how information is being processed inside.

How shall we proceed, then, in the absence of a suitable, objective vocabulary? We will muddle through by using the language of unambiguous definitions and subjective experience, but with the implicit understanding that when we use words like *anger* or *happiness*, we will agree that we mean, not a *thing*, but "the mental state that we subjectively experience as anger or happiness."

My "explanations" of these experiences will consist mostly of *plausible mechanical analogies*. These analogies may or may not model how the brain actually produces these experiences, but I do not regard this as important. What is important is that a plausible mechanical model *can be* constructed.

Once a logical structure for our own thinking is spelled out in detail, creating a physical realization of it by reverse engineering would seem just a problem in technology. But the logical structure of our own thinking patterns is as opaque to us as ever! Early optimism about understanding human thought from an information-processing point of view has been replaced with more restrained language. We are beginning to understand how deeply intertwined our thought patterns are with the accidents of our biological and social evolution and our environmental experiences. We may find it impractical, even undesirable, to replicate in detail all the human mental functions that nature took eons to develop. To be sure, many of nature's steps along the evolutionary trail are responsible for behaviors that cause us a great deal of trouble.

The study of intelligent behavior will therefore likely diverge along two different paths: One will explore the origins and structure of human thinking patterns from the point of view of the environment in which they evolved and the genetic purpose they served. This path has been called *evolutionary psychology*. The other path will pursue the evolution of intelligence less and less like our own, with utterly different awareness, priorities, and goals. As James Hogan put it in his book, *Mind Matters*, "Evolution produced flying things that have feathers and chirp, but engineering a flying machine gets you a 747."[3]

Losing Control

The consequences of an explosion of machine intelligence reach into every nook and cranny of our social fabric. Even today, the evening news usually reports several incidents of mishaps caused by the failure of complex technologies. A preview of the kind of social problem that "intelligent" computing systems pose has been called the *software crisis*. The large and complex computer programs that control and monitor so many facets of our lives are less and less comprehensible by humans. (The Windows 2000 operating system is said to contain 40 million lines of code.) The adaptive and self-modifying ("intelligent") programs in our future will be even less comprehensible and will produce less predictable results.

The potential for widespread social and economic disruption precipitated by failures of critical computer systems can hardly be overstated. When a NORAD (North American Air Defense Command) computer mistook the rising moon for a flight of attacking Soviet missiles, nuclear war could have been the result! The failure of software designers to foresee the arrival of the year 2000 prompted a panic to avert disruption of critical systems when the new millennium arrived. To avoid having to understand and fix the offending programs, many users simply replaced them.

Such catastrophes are the natural result of underestimating the chances that our machines can fail. One "obvious" solution is more care and discipline in engineering, testing, and maintaining programs, but that is not enough. It will also be essential to build into complex programs more intelligent self-monitoring and self-correcting capabilities. We will see that such capabilities are the ingredients of a rudimentary kind of consciousness. Indeed, our own consciousness can probably be traced to the survival value of such capabilities.

Perhaps we have very little control, on the average, over the course of evolution of intelligent machines—no more control than fourteenth-century Europeans had over the progress of the Black Death. If the victims of the bubonic plague had understood sanitation and the ways that pathogens are transmitted, the effects of the plague would have been minimized. But they *didn't* understand—any more than we understand today about how rapidly and inevitably intelligence evolves and takes on new and unpredictable forms. Creating machines that we cannot completely control, and then putting them in charge of critical aspects of our lives, poses risks whose consequences we may not have the luxury of contemplating afterward.

Survival

There is still time for clear, careful, rational thinking about where we are going, where we *want* to go, what we have to gain, and at what cost. As with human cloning, we are inclined to ask such questions only when the technologies are upon us—too late to deal with them effectively. So the sooner we understand, the more control we are likely to have. But if we fail to understand and adapt to the new world in which we are about to find ourselves, we may face extinction, or at least assimilation.[4]

Our place in that world will be largely determined by the *choices we make*, both as individuals and as a species. To survive our technological adolescence and to preserve even a facade of human dignity, we may have to lose some of our self-destructive evolutionary baggage. Before we can learn to live with intelligent machines, we may first have to learn how to live with each other. If we continue dealing with each other and with twenty-first-century technology, equipped only with rigid moral, ethical, and religious beliefs that haven't changed significantly in most of the world since the Middle Ages, then the machines will surely triumph. Instead of asking what will we do with intelligent machines, we may well be asking *What will they do with us?* If we're lucky, they may keep us as pets!

Is there anything we can do to keep this creepy scenario from playing out? Calls for global awareness of the problem, and for global agreement to regulate scientific inquiry into certain fields, do not paint an optimistic picture. Only in the last century have we developed technologies that could wipe out civilization, so we have not had much practice dealing with global-scale crises. Now we are talking about turning the scientific paradigm on its head! Yet, as odd as it may sound, there are still steps that we as individuals can take that may have some collective effect.

2

What Makes
Computers So Smart?

O nce upon a time, the word *computer* described a *human being* who did arithmetic for a living. In those days, everyone agreed that computing required intelligence. Today, we are not so sure. Our electronic computers are vastly more capable than their human name-sakes—performing such advanced cognitive tasks that we commonly call them *smart machines* and *electronic brains*. Yet, in a linguistic quirk, most of us still refuse to call them *intelligent*. Although our electronic machines seem to be getting smarter and smarter, we reserve the term *intelligent* for the shrinking number of tasks that only human beings perform.

How is it that "mere machines" can perform tasks so complex that we compare them with human brains? What goes on inside their silicon circuitry that seems so much like thinking? Are we ourselves merely organic machines executing programs coded in our genes and shaped by our environment? Is human thought just a fancy kind of computing? If so, will machines eventually get so smart that they will be able to do anything a person can do? If not, where is the gap that no machine will ever be able to bridge? Are there any limits to what machines whose capacities are not confined by organic vessels might be able to do?

In Our Image

The main reason we call some machines smart is that we've created them in our own image, that is, to perform the same kinds of mental tasks that we do—only faster and more accurately. Early humans first created mechanical computers (like Stonehenge) to keep track of the time, the seasons, the positions of heavenly bodies, and the rise and fall of the tides. These machines served as a kind of memory aid that followed the cycles of nature and helped the first farmers make decisions about planting and sailors to navigate the seas. After the invention of currency, mechanical calculators like the abacus made arithmetic and trade easier and faster. A third human function that machines help us with is making decisions. *Make decisions? Since when can machines make decisions?* The first crude examples may have been tossing a coin or throwing dice to let supposedly random forces (or gods) decide future courses of action. More sophisticated devices, like a tail on a kite, a governor on a steam engine, or the thermostat on your furnace, make second-to-second decisions that automatically regulate and stabilize machines so that humans don't have to continuously adjust them.

So mechanisms have been around for centuries to help people *remember*, *calculate*, and *make decisions*. If you think about it, you will recognize that most computers in use today still serve these three basic functions, just in more elaborate combinations.

Today, we continue to develop machines to replicate more advanced human abilities—sensors that interact with their surroundings, speech, vision, language, and locomotion—even the ability to learn from their experiences to better achieve their goals. For example, many kinds of *robotic pets* now commercially available interact with their world, make seemingly autonomous decisions about their next actions, and communicate with people in ways that seem to express emotions like joy, anger, and anxiety.

Other learning machines emulate the part of human behavior that amasses knowledge and keeps track of the relations between things. These *expert systems* often know more about their specialty than do their human counterparts. Expert systems already compile and extend the experience of the world's specialists by dispensing medical diagnoses, legal analyses, and psychological advice, and by forecasting the weather, locating natural resources, and designing vehicles, structures—and even new computers!

Consequently, our efforts to supplement and enhance human intelligence have already paid off with machines that actually *exceed* human thinking capacity in many specialized areas: calculating, decision making, remembering, learning, creating, making plans, pursuing goals, and—who knows?—perhaps even enjoying it! A reasonable extrapolation of present capabilities would predict the connection and cooperation of computing *systems* that will increase the *breadth* of machine intelligence as well. Joined by networks like the Internet, machines might even combine and integrate specialized capabilities and eventually evolve their own communal and social rules akin to ethics and morality.

Machines are just beginning to learn how to modify their own programs, even to reproduce and improve their own kind, but in a way far superior to the way biological organisms do. Machines can pass along *everything they have learned* to succeeding generations. This ability gives machines a tremendous evolutionary advantage over humans, who do this in a crude way by recording knowledge and experiences outside their brains in external media, like books or the Internet. Only fragments of this knowledge and experience are actually absorbed by succeeding generations.

Soon, we can expect machines to exhibit emotions and consciousness (more about them in Chapters 7 and 8) and to advance to new kinds of intelligent behavior that we are incapable of understanding. Once machines realize that they need not be limited to the kinds of tasks that humans can do—that this is not the essence of what makes them so smart—then they will have transcended their original purpose of serving man. They will have advanced beyond mere tools of man to *independent agents*.

Divide and Conquer

Even though we call computers smart, or even intelligent, because they emulate human accomplishments, such labels tell us nothing about how smart machines, or human minds, for that matter, actually carry out their cognitive tasks. The key to performing a complex job like recognizing a song, proving a theorem, or playing chess lies in breaking it down into many simpler steps, or instructions.

The simplest imaginable step is opening and closing a switch. Information-theory pioneer Claude Shannon showed that you can perform any logical task, no matter how complex, using an arrangement of off-on switches. A brain performs its basic switching functions at the level of a neuron. A digital computer carries out its elemental switching at the level of a transistor. In the future, instructions may be carried out more efficiently by switching beams of light or quantum states. But the nature of the machine that carries out the steps is unimportant. Advances in storage capacity and processing speed will simply keep up with the demands of more complex tasks. What makes machines so smart is that they organize and manipulate information that is represented symbolically by complex switching networks.

In its most general terms, then, any computer, no matter how simple or complex, is no more and no less than *a switching network that takes one set of symbols and transforms them into another set of symbols*. The transformation can be as simple and obvious as the one called "+," which transforms the symbols "2" and "3" into another symbol, "5." An instruction that makes a simple decision would be something like: *If the number in location A is positive, then execute instruction B; otherwise execute instruction C.* Using this kind of if-then rule, a program can "bring to life" either of two different courses of action, *B* or *C,* depending on what value happens to be in *A* at the time. It is this simple logic that allows a program to adapt to the situation at hand. By combining many steps of this sort, it is possible to create, for example, the moderately complex transformation of the arrangements of chess pieces on a board into a strategy for winning this complex game—or perhaps the still incomprehensible process by which the human brain transforms an immense variety of sensory input into the wonderfully complex and adaptive tapestry of human behavior.

Turing Point

In the 1940s, people like British mathematician Alan Turing and John von Neumann set off an exponential growth of computing power when they conceived the idea of a *universal computer.* Their idea was to build a general-purpose computer that manipulates symbols according to a sequence of instructions called a *program,* which could be changed to perform different tasks. In other words, you don't have to build an entirely new machine to perform each desired function. You merely *change the program* to a new set of instructions appropriate to the function you want it to perform.

(This division of labor between *hardware* and *software* was not entirely new. Back in 1804, the Jacquard loom had used different sets of punched cards to automatically program weaving machines to create cloth with any pattern. Charles Babbage and Herman Hollerith later used similar punched cards to feed data into their mechanical computers. Ada Byron, also known as Lady Lovelace, is generally credited with writing the first computer program in 1843 for Babbage's *analytical engine,* which was never actually completed.)

Since the real intelligence would lie in its program, the hardware required to create Turing's universal computer could be very simple but flexible. He imagined a simple realization that worked by writing and changing binary digits (zeros and ones, for example) on some storage medium, such as a moving paper tape, and manipulating the arrangement of the binary digits on the tape according to programmed rules. He also imagined that the *program itself* could be encoded as binary digits on the same storage medium.

But the really amazing thing that Turing proved mathematically was that such a programmable machine could, with sufficient storage capacity and speed, *fully emulate the function of any other machine*, including, he thought, the human brain, which he supposed to be just another kind of machine! But wait! Was Turing saying that everything a human mind is capable of can be reduced to computations? That manipulating bits of information is somehow the same as thinking? That machines can somehow learn to be smarter than the humans who create their programs? That they could someday function truly independently of humans? If Turing was right (and no one has yet proven otherwise), then only engineering limitations on its speed and capacity prevent a universal computing machine from accomplishing any physically possible task!

A Symphony of Ones and Zeros

Most computers in use today are the direct descendants of Turing's concept of a universal computer. All the computer functions that we take for granted today—word processing, e-mail, the World Wide Web—are performed by programs that simulate these functions on general-purpose digital computers.

When you dig down inside a modern digital computer, to the level at which information is being processed, you find an electronic version of Turing's paper-tape machine. The computer manipulates strings of zeros and ones, or more precisely, switches electrical voltages between two states representing zeros and ones. There are no pictures, no music, no books, no spreadsheets, no Web sites inside your PC—just electrical circuits for manipulating arrays of binary digits, or *bits*. Anything a digital computer does can be reduced to a huge number of simple operations on these binary off-on voltages.

What a remarkable discovery—that we can reduce any kind of information to binary digits! Every sensation, everything we see, hear, say, write, even taste, smell, and touch, can all be reduced to a collection of ones and zeros. So can all the works of Shakespeare, the music of Mozart, and every movie ever made. With the completion of the Human Genome Project, we can now represent the sequence of nucleotides in our DNA—the instructions for making a human being—as a symphony of ones and zeros!

The Limits of Divide and Conquer

It is natural to ask at this point whether *everything* we experience can really be broken down into a bunch of binary computations. Can we really reduce to step-by-step instructions the beauty of a sunset, the smell of a rose, or the love of a mother for her child? The premise upon which AI is based is that *any task whose details we can describe precisely can be done by a machine.* Some AI researchers believe that all that is required to reproduce *any* behavior on a machine is to precisely describe its *logical structure.* If so, then we would be able to build a feeling machine by precisely spelling out the structure of certain emotions, or a conscious machine by telling it exactly what consciousness means in minute detail.

But others say that there are fundamental limits to what can be logically analyzed in this divide-and-conquer way. The most glaring failure of a *reductionist* point of view, they say, is its inability to model the fantastically complex behaviors of life-forms, including those of human beings. No matter how much you chop up a feeling or an idea, you will never find the basic units of thought or emotions. The idea of a nonmaterial *soul* or *spirit* was invented to fill this gap.

It's easy to attack mechanistic models of behavior and, indeed, scientific explanations in general by labeling them *reductionist* when, in fact, they are not. Examining things by breaking them up into manageable pieces is certainly a traditional tool of science, but the essence of scientific understanding consists of more than analyzing pieces. How, for example, could you understand the way a human being functions simply by studying its individual organs (or even cells) in isolation? The answer, of course, is that you can't. Breaking down a complex system into its elemental pieces is only the *first step* toward understanding how it works; the next step is understanding how all those pieces *interact, fit together, and cooperate.*

A classic argument that science is not merely reductionism cites the properties of water: You can't understand water simply by examining the properties of its component atoms, hydrogen and oxygen, alone. To predict the physical and chemical properties of water, you have to understand the physics of how hydrogen and oxygen atoms interact.

A complicated task like writing this book can be broken down into chapters, sections, sentences, and words. But combining these parts willy-nilly won't do the job. The way the parts are put together requires some kind of *vision* about how words, sentences, paragraphs, and chapters combine to form a coherent whole.

In the fuss over the role of tire separations in sport utility vehicle (SUV) rollovers in the late 1990s and later, investigators tried to blame either a specific brand of tire or a specific model of SUV. The root cause, however, was most likely a flawed *combination* of the two.

The point here is that a *holistic* view is an essential part of all knowledge. If you want to tell someone how to perform a complex task, then the instructions will usually consist of more than a linear list of simple steps. Purposeful behavior, no matter how intricate, subtle, or complex, must be understood as a large collection, or web, of relatively simple operations *acting together*. Acting together means that the steps must be *organized and coordinated* so that they work in concert, like the instruments of an orchestra, to produce the desired result. Complicated tasks may require many subtasks to be completed in parallel and their results integrated. Many jobs require complex branching or recursion (taking the results of subtasks and feeding them back to the beginning again and again). But monitoring, coordinating, and integrating large numbers of subtasks is the job of an *executive* that takes a holistic view and checks to be sure that the result is the desired one. The top executive, who may be a human or another program, is said to be *smart enough to know* what the overall problem is and to organize the steps required to solve it. This purposeful organization of tasks and subtasks is what makes computers so smart—and is indistinguishable from what we call *thinking*. It would seem, then, that there should be no limit to the number of tasks, each with its own executive, that can be organized into hierarchies, like human corporations, and combined to perform ever more complex tasks, including tasks that we would call purposeful or intelligent.

There are two ways that such purposeful organizations can come about. One is the traditional top-down method of design, in which a functional requirement is first defined and then the necessary systems and subsystems are created and integrated to produce the desired outcome. This is how human beings are used to thinking about and solving problems. An autonomous planetary explorer designed by the National Aeronautics and Space Administration (NASA) would be an example of a top-down-designed "smart system."

A second way to create complex organizations is called the bottom-up approach.[1] It begins with simple pieces that self-assemble in a random, trial-and-error fashion. Some combinations persist that are able to make copies of themselves or to further assemble into more complex combinations. Environmental forces determine which combinations survive or die. This is the mechanism we call *natural selection*, a much more time-consuming process whose results are highly optimized but essentially unpredictable.

One limit, in practice, to the power of such intelligent systems, whether built or grown, is the sophistication of the *error checking* and *quality control* that are required to keep such complex hierarchies from collapsing of their own weight. Self-maintenance, as we will see in Chapter 7, is crucial to the survival of any "smart" mechanism, be it machine or human.

3

What Do You Mean, Smarter than Us?

[M]achine brains can be much faster in the way they oper-
ate or make decisions, they can be much more accurate,
more reliable, keep performing at the same level for a long
time, deal with numerous things at (roughly) the same time,
consider many different possibilities very quickly, remem-
ber facts accurately, perform complex mathematics rapidly,
learn much more quickly, and so on.

KEVIN WARWICK

Professor Warwick concisely sums up what we usually mean when we say that machines might be getting smarter than us.[1] The list of machine accomplishments encroaches daily into the domain we used to reserve for human achievements: Consumers have smart phones, smart cameras, smart credit cards, and smart cars. Computers play championship-level chess, bridge, checkers, and backgammon. They translate technical manuals, pilot automobiles across the country, prove mathematical theorems, and implement stock-trading strategies. Our robots explore the ocean's depths and other planets, clean up hazardous waste, and manufacture goods, including other computers! We make smart weapons to fight our wars—but *we* are not yet smart enough to stop fighting them!

Does this list really mean that computers are getting smarter than us? Which of these accomplishments would we call intelligent? The answer seems to depend more on personal perspectives and what is socially acceptable than on any objective definitions of these words. Many of us refuse to label any machine, no matter what it does, as truly intelligent. Others glibly use the term to describe any apparently purposeful behavior, be it machine, human, or other animal. Are there any truly objective indicators of intelligent behavior? How might we recognize it when we see it? We will revisit these questions many times throughout this book.

The Turing Test

There have been many attempts to avoid precisely defining intelligence by inventing "objective" tests for detecting it. The most famous of these is called the *Turing test*, an adaptation of an "imitation game" suggested by Alan Turing in 1950.[2] (Turing is also famous for breaking the German codes in World War II.) The idea of this thought experiment (which, by the way, Turing never intended to be used as an intelligence detector) is that a blind interview would be set up, in which a human interviewer would converse, by typing on a terminal, with an unseen something or someone in the next room. Any subject matter is allowed in the conversation. Various humans and machines are given a turn conversing with the interviewer, whose job is to guess, in each case, whether he or she is communicating with another human or a machine. (Perhaps the modern equivalent would be trying to determine whether your e-mail pen pal is a machine or a person.) Any subject that succeeds in fooling the interviewer after some suitable time is said to be intelligent.

Unfortunately, the Turing test merely replaces the semantic issues with questions about which aspects of intelligence the test is allowed to cover—that is, the rules about the scope of the test. The rules seem, on the surface, to require only that the subject understand language and conversational conventions, be able to reason, and have command of a suitable collection of facts and information. But a full-spectrum Turing test would also have to allow questions designed to reveal aesthetics, emotions, creativity, moral values, common sense, and so forth. Because this is such a tall order, many people—like philosopher Daniel C. Dennett—believe that a machine that could pass an unrestricted, full-spectrum Turing test would have to be called intelligent.[3] In other words, it is able to think, by anyone's definition of the word. Some say it should therefore be accorded civil rights.

The trouble is that a truly unrestricted or full-spectrum Turing test is impossible to construct and would take an infinite amount of time to administer. So we are left with these questions: How comprehensive must a Turing test be, so that a subject passing it is said to think or should be accorded civil rights? Shouldn't it be possible to devise an abridged test to decide these practical issues? Because there is no such thing as *the* Turing test—there are only grades of difficulty—these questions force us back into the impossible task of precisely defining intelligence (or moral responsibility, or whatever) in an objective way that is amenable to testing. This is hardly progress.

Nevertheless, volumes have been written about the philosophical and practical relevance of the Turing test. A lot of programmers have tried to design conversational programs to pass various abridged versions of the test. These efforts have been encouraged by a much-criticized $100,000 prize offered by philanthropist Hugh Loebner in annual Turing-test competitions, beginning in 1991. The Loebner Prize for the "most human" computer has been won two years in a row (2000 and 2001) by a program called ALICE, with which you can converse on the Web. Lots of amusing anecdotes and transcripts have arisen from these and similar competitions. For example, one human taking part in such a test was judged to be a machine because she produced coherent paragraphs of grammatically correct and informative prose—apparently something the interviewer felt that no human was likely to do!

Another problem with the Turing test is that any rules you can think of are clearly human-centric, that is, they focus on the narrow spectrum of behavior peculiar to human conversation. This is a patently anthropocentric definition of intelligence. The test does not detect intelligence in general—it is strongly biased in favor of the human variety. Programs like ALICE are said to be "convincingly lifelike," not because they are so intelligent, but because people are so dumb.

Even if we could agree on some abridged version of the test that would be a *sufficient condition* for claiming that a machine can think, would it also provide a *necessary* condition? In other words, must a subject pass the test to prove that it is intelligent? Shall we deny it civil rights if it can't? We can imagine many humans (like the aforementioned literate woman, as well as uneducated people) who could not pass it. A superintelligent extraterrestrial would probably fail as well. And couldn't there be a computer so sophisticated that we would not hesitate to say it thinks, yet which lacks the specific language skills required to pass the Turing test?

Writing programs to fool an interviewer in a Turing test is a useful AI exercise that offers insights into how people think and converse. But its practical significance as a test for intelligent behavior has been exaggerated far beyond the simple mental exercise that Turing proposed. Emulating human conversation certainly has useful engineering applications, but it contributes no more to understanding the general principles of intelligence than knowing how a sparrow flies makes you an aeronautical engineer.

One lesson to be learned from the abuses of the Turing test is that we are not likely to make great strides in the science or engineering of artificial intelligence merely by imitating human thinking. First, the difficulties of doing so are monumental, but more importantly, why limit ourselves to creating humanlike intelligence? A broader perspective allows us to think about entirely different kinds of intelligence than the human kind. We will consider such new directions further in Chapter 6.

Intelligence Is Multifaceted

A machine that passes a reasonable Turing test must meet or surpass human conversational ability, but this is in fact just *one of many* yardsticks that might be used to compare human and machine intelligence. We can also build machines that are smarter than us in other useful ways, but that lack the specific abilities required to pass the Turing test. A pocket calculator would score very high against a human in the one dimension of computational ability, but it would fall miserably short by virtually any other yardstick. A chess-playing machine would score very high relative to most human chess players, but it would not be a great conversationalist.

Any claim to be able to measure whether one thing is smarter than another implies that the "smartness scale" is one-dimensional. It makes sense to ask if I am taller than you, because tallness is a one-dimensional property. We can quickly answer the question with a tape measure. But if we ask if I am smarter than you, how would we measure that? An IQ test would give us each a single number, but such tests contain notorious biases and omissions.

Most of us understand that smartness or intelligence is multifaceted—that there are lots of measures of intelligence that are more or less independent. The creators of college entrance exams appreciate this in a crude way by testing separately one's math and verbal skills—just two dimensions of intelligence. But in addition to math and language skills, we have other kinds of intelligence, such as pattern recognition, kinesthetic ability, musical talent, knowledge base, goal-setting, and locomotion, to name only a few. So if we want to compare machine (or human) *A* with machine (or human)

B, we end up with (in mathematical lingo) a multidimensional vector that gives a relative score for each intelligence measure. The idea of "more or less intelligent" is such an oversimplification of a complex relationship that any scheme that claims to measure overall intelligence should be met with considerable skepticism.

We see now that the question of machines being smarter than people is not as simple as comparing the intelligence of each with a simple ruler. Machines are, in fact, *already* more intelligent than we are in many specialized domains! With the number of these narrow domains increasing all the time, it seems likely that the gaps between these domains will fill in, so that the *breadth* of machine intelligence will also continue to increase. As machines become more flexible and adaptable, what is to stop their overall intelligence from approaching and then exceeding that of humans?

Do We Want Machines That Are Smarter than Us?

On my home PC, I use a word processor that has a grammar-check feature. It's supposed to look through my writing and find not only grammatical errors, but also awkward phrases and constructions, as well as jargon. When I use this feature, it often flags words and phrases that might be incorrect or awkward in other contexts, but when I agree to the correction, the program often changes the whole meaning of what I wanted to say. It often turns out that the way I said it is (in my opinion) the best way to say what I wanted to say. So what good is a machine that's "smarter" than me, if it's going to change the meaning of my writing? I never use such grammar-check features anymore.

The grammar-check function is an example of a computer program that is *supposed* to be smarter than most people (about grammar and usage, at least), but is obviously only a very primitive version of what it claims to be. It also suggests how complex the task of creating coherent conversation is. Checking my spelling is one thing, but figuring out what I *mean* is quite another. And even if it claimed to know what I mean, would I trust it to do so? And how would I feel if a machine actually seemed to know what I wanted to say and offered to express it better than I could?

This is just a preview of the kinds of control issues that smart machines raise. Sooner or later, we will have to face the disturbing question: *Do we want machines that we cannot control?* Try to imagine some kind of benign machine that is vastly more intelligent than us, yet remains under our control. Data, the android in *Star Trek*, is such a machine, but he presents a troublesome paradox: If he is so intellectually superior, how come he allows himself to serve under human beings? If he had emotions, he would surely cringe every time he has to obey a stupid human order! It seems much more likely that the smarter machines become, the more control we will relinquish to them. If we don't think this through before it happens, we are likely to lose control so gradually and painlessly that we will never notice.

In the coming decades, we will build machines and networks of machines that excel (i.e., are "smarter") in some significant, but not all, aspects of human intelligence. It is highly unlikely, however, that there will be machines indistinguishable in all respects from humans anytime soon, if ever. (There would simply be no point in doing so.) So rather than measuring a machine's overall intelligence against a human's, we will find it more useful to compare the speed, cost, efficiency, and accuracy of a *given task* when performed by a human, with the results of assigning a machine to perform the same task. We readily concede, for example, that an electronic calculator far outperforms humans at arithmetic and that a computer keeps much better track of the documents in a library than a human could. Although the number of tasks that machines can do better than us will steadily increase, we will not notice any particular milestone in time when we will say that machines have become smarter than we are.

Soon, we will find that intelligent machines are capable of far more than just doing things better than humans do. Machines that are able to learn and innovate and modify themselves will begin to exhibit new forms of intelligence that we will barely comprehend. How can you tell if a machine that's a lot smarter than you is working the way it's supposed to? This is the point at which we should really begin to worry about our future as the dominant life-form on the planet.

Intelligence Is As Intelligence Does

The purpose of asking the question *Could machines ever be smarter than us?* was not to philosophize about the meaning of the question, but to help us deal with the practical, everyday consequences of *any* degree of machine intelligence. It is not necessary for a machine to outperform humans in every way (that is, to be smarter, overall, than us) for its actions to command our respect and for their consequences to raise important moral and ethical questions. Suppose, for example, that an expert system consistently outperforms medical professionals in diagnosing and treating certain illnesses. It would be foolish to insist that the expert system also be able to compose music before we would concede that its decisions have important medical, social, legal, ethical, and moral implications. Arguing "It's only a machine!" will be irrelevant in considering these implications.

If we insist on measuring and comparing intelligent behaviors, whether those of humans, machines, animals, or extraterrestrial beings, perhaps it would be more useful to look at their *accomplishments as a species*. In other words, *intelligence is as intelligence does*. We cannot yet assess the intelligence of machines *as a species*, because nothing like a coherent machine society yet exists that behaves like an organic species. But this is just a matter of time. We already have many kinds of systems in which machines pass information and instructions to each other. The thermostat that controls your furnace is a simple example; the integrated systems in your car or in a spacecraft are more complicated communities that process sensory inputs, make decisions, and pursue goals virtually autonomously. When machines are able to set and pursue goals of sufficiently high level, we will be able to speak of and assess their accomplishments.

Smart Machines in Our Future

At the moment, we seem to be largely in control of the machines. (It may not seem so when you are stuck in an automated customer-service telephone system, or when you receive recorded telephone solicitations.) Yet in many ways, this control is slipping away as we leave more and more decisions to them. We like to think that we will always reserve the crucial decisions to ourselves, but this is probably wishful thinking. It is far easier to allow the machines to take over, not just to perform the physical drudgery, but also to make more and more of the technological decisions that we are no longer competent to make. What will life be like for us when we no longer make the major decisions about our world?

Again, we have to ask: *Is this what we want?* We—and I mean each of us—have a *choice* about how we respond to the proliferation of intelligent machines now upon us. If we see events unfolding in ways that are not to our liking, there is still time to act. Simply going with the flow and blindly embracing every new technology in the hope that everything will turn out OK is no longer an intelligent choice. If we simply accept machine idiocy, if we meekly conform to a world shaped by the needs of machines instead of the needs of humans, then the days to our extinction are surely numbered. If, on the other hand, we actively demand that our machines enrich our lives, and if we take part in the process, then it is more likely to turn out that way.

The process of evolving machine intelligence will necessarily be littered with the corpses of failed attempts. This is how evolution works. It is *our* job to make sure that what survives is only the machines that make our lives more productive and satisfying—and to decide what *productive* and *satisfying* mean. If we choose wisely, then science and technology will serve us well; if we do not, then we will be its slaves. The history of technology tells us that we always make a combination of wise and foolish choices, and there is no reason to expect this to change. But we may tip the balance toward wiser choices if we make sure they are guided by human needs and values. So perhaps the deepest questions we are left to ponder are *What are those needs?* and *What are those values?*

Machines will more easily displace us if we allow our own intelligence to languish. How shall we compensate for losing more and more functionality to machines? An extraterrestrial observer might judge our intelligence by the uses we make of our technology. Which is smarter, a species that has figured out how to survive its technological adolescence to live in peace, or one that uses its technology to destroy itself? One that consumes all its resources and fouls its home without regard for future generations, or one that has established an equilibrium with its environment? By these measures, our extraterrestrial observer might conclude that although our machines are getting smarter, we are getting dumber.

4

Machines Who Think

The idea that your essence is software seems a very small step from the view that your essence is spirit.

HANS MORAVEC

The question *Can machines think?* sounds at first like an intriguing one—until you realize that the answer depends entirely on how you define *machine, think,* and even *can.*[1] By suitably defining these words, you can get any answer you want. For this reason, Alan Turing, in his 1950 article, dismissed the question as "too meaningless to deserve discussion."[2] Yet in our daily speech, we often attribute thinking to machines. We are used to saying things like "My PC *thinks* the disk is full," or "It *expects* the file to be in this format" and "It *wants* a disk in drive *A*." When I have a fire going in my living room, I sometimes say that the thermostat *thinks* it's warm enough and doesn't turn the furnace on, even though it's freezing in my bedroom. Are these merely anthropomorphisms, or are we attributing rudimentary thinking powers to those machines? In what sense might a machine be said to think?[3]

Most of the philosophical verbiage spent addressing such questions can be bypassed if we agree that there is no sharp boundary or test (such as the Turing test) that separates thinking from nonthinking entities, that is, that intelligence has no precise definition. Let us suppose instead that thinking and intelligence, as we commonly use these terms, are not only a matter of degree, but are multidimensional as well. The kind of inner experience and conscious awareness (Chapter 7) that humans have—or think they have—occupies only one point (or perhaps 6 billion points) in "intelligence space." Elsewhere in this space lie the cognitive powers of animals and machines. Each possesses cognitive abilities uniquely suited to the tasks it must perform to survive. Just as a dog's thinking differs from a cat's, a machine's differs from a human's.

Early AI visions of thinking machines were influenced by the humanoid robots, or androids, of science fiction. Today, that goal of reproducing a complete set of human sensory and cognitive abilities in a machine is regarded as not only unduly optimistic, but wildly off target. Those early visions not only spectacularly miscalculated the enormity of the gap between computing machines and even the simplest biological systems; they also missed the point. First, creating a machine that would interact with the world in a fully human-like way would have to duplicate the entire human experience of being born and raised in that world. *But we already know how to do that!* Furthermore, there are certain aspects of human behavior that we would be ill advised to replicate.

Instead, AI research and development have focused on machines that perform many different useful and specialized tasks that would be called superintelligent if they were performed by people. In those special roles, then, machines are already "smarter than us." Each occupies its own region of the multidimensional cognitive space, with some overlaps between machine and human cognitive abilities. Those regions will continue to expand.

Not If, but When

In 1965, Gordon Moore, cofounder of Intel, predicted that computer processing power would double every eighteen months. If anything, Moore's Law, as it has come to be called, has proven to be an underestimate. Because perceived physical limitations on processing speed and information storage regularly fall to advances in materials science, developments on the hardware side of computing still seem to be growing at Moore's exponential pace. Robotics scientist Hans Moravec estimates that the computing power of the human brain is about 10^{14} operations per second, and its storage capacity is about 10^{14} bytes.[4] At the present rate of growth, the power and memory of computers that are widely available will exceed these numbers between 2020 and 2030. So today, it is common to read predictions that, thirty to fifty years from now, we will have machines that exceed human intelligence.

These predictions are not that scary to most people, who (quite correctly) do not equate computing power with intelligence. Why not? If you create a computer whose speed and memory capacity exceed the brain's, won't it automatically begin to think? Anyone who has ever used a computer knows that the answer is no. Someone has to write the *software* for human thinking, which in turn requires that someone has to understand completely how human thinking works—including consciousness and emotions! This is why predictions about machines performing at levels approaching the capabilities of the human brain consistently fall short of their mark.

In the 1950s, there was a lot of hoopla in the AI community about the imminent ability of computers to converse in natural language, but that goal has still not been achieved. What went wrong? The flaw in these and other predictions now seems clear: Although the hardware has been adequate, the predictions grossly underestimated the AI community's ability to write software that encompasses the complexity and subtlety of language. Moore's Law notwithstanding, good software is always harder to write (and maintain) than anyone thinks!

Predicting the future by extrapolating present trends is always a risky business. During the 1940s, popular science magazines predicted that in the postwar future, everyone would be commuting to work by private plane or in something called an autogyro. In the 1960s, sociologists who examined trends in the age of retirement predicted that by 2000 most of us would be working less and retiring by age fifty. Anyone who invested in the dot-com boom of the late 1990s painfully appreciates the risks of extrapolation. Now, with twenty-twenty hindsight, we can easily see what all these predictions failed to take into account.

What those who predict that computers will soon exceed human intelligence fail to take into account is that the ultimate goal of AI is *not* to create machines that think and act in all respects like humans. A more likely future will contain intelligent machines whose forms and abilities are quite unhuman, and which will occupy more and more advanced parts of cognitive space. The real computer revolution will begin when we stop making machines that emulate aspects of human thought and enable them to develop new thought patterns of their own. As for replacing us, that has already begun.

How Can a Computer Be Smarter than the Humans Who Program It?

A persistent objection to the idea of machine intelligence is that a machine can never do more than execute the set of instructions it is given. And since those instructions are (so far) always created by humans, or even large teams of humans, no machine can ever be smarter than its creators. Even if a machine is programmed to reproduce some aspects of human thinking, it is just mimicking, or following a script. It can never, the argument goes, create anything new or original, only manipulate bits and bytes according to rules rigidly prescribed by its program, no matter how elaborate that program might be. Is it true that nothing original can be made entirely from a set of instructions? Is this a reasonable objection or just rhetorical trickery?

First of all, the argument contains a logical fallacy called a false unstated premise. It goes like this:

1. Machines follow only instructions (programs) written by people.
2. But such instructions are limited to the knowledge and experience of the people who write them.
3. Therefore, machines that carry out those instructions are limited by the ignorance of their programmers.

Statement 2 is the unstated assumption. You can easily show that it is false by thinking of instructions that require a machine to use knowledge and experience not directly available to its programmers. For example, ask your electronic calculator to compute the square root of 321. It uses an algorithm, devised by a mathematician, to compute the square root of any number you can think of, but it is unlikely that the programmer ever knew that the answer for this case is 17.916. . . . In other words, a calculator possesses mathematical "skills" that its programmers do not have.

It is easy to find other examples, besides arithmetic, in which a machine comes up with information unknown to its programmer—searching the Web, matching fingerprints with a national database, and superhuman expertise in areas like medical diagnosis. If we fill a computer's memory with the contents of the Library of Congress, suitably indexed, cross-linked, and cross-referenced, wouldn't such a machine have a command of facts and information far exceeding its programmer's?

Learning Machines

Suppose we give a machine this even more elegant instruction: *Go out into the world and learn as much as you can.*

A machine with suitable sensors, mobility, sufficient memory capacity, and the processing power to carry out such an instruction could learn things by using neural nets to record and reinforce information it acquires about the world. Eventually, after perhaps many years of "experience," our learning machine would have acquired a working knowledge of how to function in the world, including language and what we call common sense. Furthermore, it would have acquired significant skills and knowledge not known to its programmers. In other words, it uses what it learns to change its own program! It might develop its own symbolic shortcuts, similar to our language, to efficiently store and organize information and even to communicate with other machines. A machine that continuously updates its own program would behave in autonomous and unpredictable ways (Chapter 5). In principle, there is no limit to the amount of knowledge that our machine might acquire in this way—the same way people do, but without those annoying biological limitations!

Yes, But a Computer Can Never _____!

When told how machines are getting more capable all the time, many people say, "That's fine, but a computer will never be able to (fill in the blank)." The task in the blank is usually something like *feel emotions, enjoy a sunset, fall in love, be creative, be conscious, believe in God*. No proof is ever given for these assertions. The reasoning—if one can call it that—is basically this: *I've never seen a machine that does these things; therefore, no machine ever could do these things.*

One counterargument is that the list of things a machine "can't do" keeps shrinking. It used to include things like write a symphony or play a decent game of chess—supposedly human abilities now shared with machines. One by one, qualities that we cherish as being uniquely human are being learned by machines. Many say *imitated* by machines, but they miss the point. Much of what humans call learning might be called imitation as well. What they seem to be saying is that a computer can never *be human*. Well, yes! In the vast multidimensional cognitive space, why should we choose to make machines that occupy the exact point(s) that humans (or any other biological creatures, for that matter) already occupy?

Ask Me Anything

I think it is much more interesting and useful to speculate on what computers *can* do, rather than on unprovable assertions about what they *can't* do. Practical forms of AI already exist in the form of programs called *expert systems*. An expert system is an organized collection of knowledge about some narrow domain, combined with an interface that allows the user to interact with that *knowledge base* in a systematic and practically useful way. Expert systems often use decision trees that lead the user through a process like diagnosing illnesses or exploring for minerals. The knowledge base is constructed from statistical information gathered from a large sample of cases and by picking the brains of many human experts. As a result, some expert systems outperform individual human experts.

The most familiar, successful, and widely used expert system that I know of is TurboTax, the PC software that leads you through the preparation of U.S. income tax returns. This truly amazing program fully justifies the label *expert system*, since it appears to be able to correctly answer, in usable language, virtually any practical question about the arcane U.S. Tax Code that the individual taxpayer might encounter. It is also amazing because it is "friendly" and adaptive enough to make it a wildly successful commercial product. (No, I don't own any stock in Intuit, the company that makes TurboTax.) If anyone ever asks you if AI will ever have any practical applications, TurboTax is a good example.

Early rule-based expert systems were developed to assist in the diagnosis and treatment of medical conditions, and this application continues today. MYCIN is a famous program written to see if an expert system could help doctors diagnose bacterial infections in the blood according to the symptoms and test results presented. It interacts with a medical technician by asking a series of questions. This is the sort of "conversation" a technician might have with MYCIN:[5]

> MYCIN: Please enter information about the patient.
>
> TECHNICIAN: (enters relevant data, such as name, age, sex, weight, race)
>
> M: Are there any positive cultures obtained from Sally?
>
> T: No.
>
> M: Are there any pending cultures of Sally?
>
> T: Yes. (enters type and results)
>
> . . .
>
> M: Has Sally recently had symptoms of persistent headache or other abnormal neurological symptoms (dizziness, lethargy, etc.)?
>
> T: Yes.
>
> M: Has Sally recently had objective evidence of abnormal neurological signs (nuchal rigidity, coma, seizures, etc.) documented by physician observation or examination?
>
> T: Yes.
>
> M: The CSF cultures will be considered to be associated with MENINGITIS.
>
> . . .
>
> M: Is the meningitis a hospital-acquired infection?
>
> T: No.
>
> M: Is Sally's clinical history consistent with EPIGLOTTITIS?
>
> T: No.
>
> M: Is Sally a burn patient?
>
> . . . and so on.

When MYCIN completes its queries, it offers a diagnosis and recommends a specific therapy, including medication doses customized for this patient's needs.

The successors of MYCIN seem potentially capable of revolutionizing medical care. Other well-known medical expert systems include CASNET for glaucoma, INTERNIST for internal medicine, VM for intensive care, PUFF for respiratory conditions, and ONCOCIN for cancer. Similar expert systems are now used by the medical profession, both to aid actual diagnoses and for training purposes. Most systems could use some enhancements, like knowing about details of the patient's history and the medical care he or she has already received. Consequently, no patient is yet ready to entrust his or her care entirely to such a system. But it seems only a matter of time, if it is not yet so, until these systems will consistently outperform their human counterparts. The holographic doctor on *Star Trek: Voyager* perhaps portrays the ultimate medical expert system. This prospect raises many challenging legal and ethical questions, some of which we will explore in Chapter 11.

Do Boxes Understand?

Volumes have been written debating the question of whether a machine that simply manipulates symbols can ever be said to truly think or *understand*. Most people have the subjective feeling that understanding must involve something more than manipulating symbols, although few of them can specify exactly what that *something more* is.

Joseph Weizenbaum designed his famous program ELIZA (also known as DOCTOR) to test the idea that a plausible human conversation could be carried on by a relatively simple interactive computer program. It does so by interpreting text typed in on a terminal and responding in the manner of a Rogerian psychotherapist. Questions and comments that the "patient" types in are scanned for words that match a small dictionary of words that psychiatric patients typically use, like *problem, help, need, mother, father,* and *family*. Then, appropriate responses are constructed, either by feeding back the submitted words in a "responsive" context or, if no matches are found, by answering with canned responses in the Rogerian style. It is perhaps a poignant statement about the craft of psychotherapy that at least one therapist supposedly commented that a slightly enhanced version of the ELIZA program would be an efficient means of treating real patients! You can now find and interact with various versions of ELIZA on the Web.

Who is to say whether ELIZA is a rather limited prototype of a thinking machine? It "listens," makes decisions, and responds in a convincing and sympathetic way—perhaps as well as some doctors do—yet it is not difficult to trip it up and elicit nonsensical responses. If you look at the amazingly simple program inside ELIZA, you would find nothing in there that thinks or understands, only string comparisons and branches in a decision tree. These days, writing such programs is a beginning exercise for AI students. Yet there have been numerous anecdotes about people interacting with ELIZA in ways that suggest they think they are dealing with a sentient (and helpful) creature. The anti-AI bunch says these people have been deluded, since this program has nothing to do with understanding, cognition, or therapy, no matter how competent and perceptive the looked-up responses seem to be.

Another classic argument against machine understanding (invented by John Searle) has been a thought experiment called the Chinese Room.[6] Suppose that instead of a computer program inside a closed room, we have a human who understands no Chinese, and that we slip pieces of paper into the room that contain only Chinese characters. The human inside dutifully looks up the input characters in a Chinese phrase book, copies down the appropriate Chinese responses, and pushes them out the slot. Because the human inside understands no Chinese whatsoever, it is "obvious" that no "understanding" occurs inside the room, even though the responses might make perfect sense to a Chinese person who provides the input and reads the output.

By making ELIZA's dictionary and the Chinese Room's phrase book more and more comprehensive, one could come arbitrarily close to reproducing very sophisticated and intelligent responses. If such a box also included information about ongoing sensory experiences accumulated over years, then its responses could not easily be distinguished from those of an intelligent human. Yet the conclusion drawn by the anti-machine-intelligence bunch is that no matter how elaborate the symbol-manipulating abilities inside a box or room, nothing is happening that anyone would call understanding. They would say that no symbol-manipulating box, devoid of understanding, could ever be called sentient.

What is easy to overlook in ELIZA and the Chinese Room, as well as in machines devised to pass the Turing test, is the presence of a kind of understanding that everyone would agree is real, once it is pointed out. The phrase tables used by ELIZA had to be made by someone who *understands* English and Rogerian counseling. And the phrase books used by the person in the Chinese Room had to be created by someone with an *understanding* of Chinese conversational syntax. Any Turing test machine is programmed with a human's ideas about what passes for human conversation. Consequently, although each box does no more than manipulate symbols, both the symbols and the rules for manipulating them are "contaminated" with understanding of the human kind.

But how could it be otherwise? As long as humans construct such symbol-manipulating machines, they necessarily impose their own rules by which the symbols are manipulated. Is this cheating? The only alternative is to require machines to independently acquire their own knowledge and experiences (including experiences with humans) and to make up their own cognitive rules. But even so-called autonomous learning systems acquire and process knowledge according to rules installed by humans.

So does the entire objection to boxes that think and understand go away if we recognize that they are all (so far) engineered from pieces of human thinking and understanding? And doesn't the same reasoning allow boxes to be *original* and *creative*, if we recognize that new ideas are simply new combinations of old ideas?

Does our difficulty with thinking machines also go away if we suppose that the definition of thinking is a *subjective* one? For example, ELIZA might be said to think to the degree that it seems to the "patient" to think, and not if it does not. Thinking would thus be not just a matter of degree, but also something we *perceive*, rather than something we can measure or test for objectively. Aren't these the same empirical criteria we apply to other people?

Machines That Reproduce

Those who believe that only living creatures can think have been known to argue (illogically) that thinking cannot be a purely mechanical function, because machines cannot do what all living creatures do—reproduce. Although reproduction and intelligence seem not directly related (neither being required for the other), reproduction is, however, part of the long-term process by which intelligence *evolves* in living creatures. In this same sense, engineering self-replicating machines would open the door not only for machines to increase their own cognitive powers virtually without limit, but also for them to develop new forms of nonorganic "life" from molecular to planetary scales.

When some people think of self-replicating machines, they visualize a robot that runs down to Radio Shack for parts and solders together a copy of itself. Others think of living cells that make exact copies of themselves by dividing in half. These kinds of reproduction might be called top-down, because you start with a finished product and want to make exact copies. This kind of reproduction is not limited to living systems. There has been some progress in building real electronic devices that repair and reproduce their own circuitry.[7] These machines "know" what the new generation is supposed to look like before they begin to reproduce. Sexual reproduction introduces a random combination of the parents' genes into the copies, but parents still know roughly what they'll get.

Another kind of reproduction (more accurately called propagation) might be described as bottom-up. It starts with a simple machine that modifies itself over and over again according to the same set of rules. The difference from top-down reproduction is that you never know what you will get until you do it. *Cellular automata*, in which the rules apply to cells and their neighbors on a regular, two-dimensional grid, produce interesting displays on digital computers. Properties of cellular automata, like *emergent complexity* and *self-organization*, are vividly illustrated by mathematician John Conway's Game of Life. It shows how simple, recursive rules can give rise to lifelike, emergent behaviors, including locomotion, eating, growing, reproduction, and dying, *even though the rules say nothing about such behaviors*. You can find many wonderfully complex, moving "creatures," or experiment with your own, at several Web sites devoted to the Game of Life.

The patterns that evolve in cellular automata are *deterministic*, meaning that if you start off with the same pattern and use the same rules, you will always get the same result, after any number of steps. But because the rules are recursive, the results are also *unpredictable*, meaning that the only way to find out what you will get is to run the "program." Cellular automata suggest to some a model for the complex behavior of insect colonies, such as ant colonies, in which each individual obeys only a simple set of instructions. Some go even further and say that cellular automata provide a new way of looking at the most basic structure of the universe.[8]

Darwin explained the evolution of living organisms as the result of *natural selection*, which is a combination of bottom-up and top-down reproduction. Individuals make copies of themselves (in top-down fashion) that contain random variations due to mutations and to gene sharing in creatures that reproduce sexually. When many such individuals compete for resources in an environment that contains both nutrients and predators, the individuals that are better adapted to that environment tend to survive and produce more offspring. So, in bottom-up fashion, the "rules" imposed by the environment continuously reshape the form of the creatures that must survive in it. This relationship is recursive, meaning that populations shape the environment, and vice versa (flowers and insects, for example), so outcomes are impossible to predict. But since the rules of individual reproduction are more like copying, organisms usually change significantly (evolve) only over many generations.

Natural selection is not confined to living organisms; it can also shape the evolution of machines. Here's an example: Imagine that you have a million identical computers that you want to succeed at investing in the stock market. Program all the computers to trade stocks in the usual way, but use a program whose decision criteria are determined by a million variable parameters, or "knob settings" that get set at the start. These settings, or rules, determine not just which particular stocks to buy. They also set buy-sell points according to various technical rules, weight different industries, look at different fundamentals, set levels of aggressiveness, determine the propensity for buy-and-hold versus active trading, and so on. The range of knob settings represents the entire spectrum of (rational and irrational) investment behavior, or "personalities."

Set each of the million computers' million knobs differently, according to some random rule that requires no intelligence on your part. Then start off these computers on their investment programs by giving each the same amount of money. After some period, say a year, give each computer a score that equals the percentage of its profit or loss for the year. For year 2, take all the computers with losing scores and replace their knob settings with settings copied from the winning computers, but introduce small, random variations in the settings. Repeat this procedure for many years. The effect is that the competitive environment of the marketplace will weed out the unfit programs and leave only those with the best investment strategies—survival of the fittest, if you will.

This is a crude example of a *genetic algorithm*, in which organisms compete to get their programs into succeeding generations—that is, to *reproduce*. An interesting feature of such a process is that the sets of rules you start out with are not decided by a human. They don't even have to make sense. As long as there is enough diversity in the rules, the marketplace itself will, after some generations, sort out the unfit, leaving only the most profitable strategies. Furthermore, the successful strategies—the results of generations of computers learning and reprogramming themselves—are very unlikely to make any sense to a human investor. This is how new kinds of thinking that are not derived from human thought processes will evolve.

Not Smarter than Me!

Fragile human egos will always resist the idea that machines could think or be as intelligent as people. Their resistance is based not on reason, but on fear of the consequences, should it come true. To them, the superiority of humans over machines is therefore beyond question. One participant in an on-line AI discussion group put it this way: "i dont want no robot thinking like me."

Although this comment may have been offered facetiously, it does seem to sum up a widely held view of artificial intelligence: It's fine if machines get smarter and smarter, as long as they know their place, which is to serve humankind and make our lives easier. But if machines ever begin to think like us or challenge our primacy in any way—in other words, if they start to get uppity—then it's time to put a stop to it!

Chess is one area in which the threat of computer intelligence is taken personally. It was once thought that a machine would be considered intelligent, and could think, if it could play a creditable game of chess at the grand master level. In 1997, an IBM computer named Deep Blue became the first to win a chess match against a world champion, Garry Kasparov. After the match, the press published a flood of face-saving letters and articles declaring that the event did not mean that computers could think, after all. Human vanity, it seems, keeps pushing the definition of thinking beyond the reach of machines.

A Pragmatic View

Are there any useful conclusions we can draw from asking the question *Can machines think?* First, thinking seems more usefully regarded as something we *perceive*, rather than something we can test or measure. We can say a machine thinks when we find it useful to do so. Whether a machine can be called a person, and whether "anyone is really home in there," are pretty much questions for philosophers to ponder. For practical purposes, they are irrelevant. What matters is whether it *acts in all respects* as though it understands. If ascribing mental processes to a machine helps us to understand how it works and to get it to do what we want it to do, then fine, go ahead and do it.

Many machines are already so complicated that it is difficult to interact with them without using mental metaphors. If your bank's cash machine refuses to give you some cash, when you know that there's enough in your account, it now seems completely natural to call up the bank and say, "Your cash machine doesn't *want* to give me any cash because it doesn't *know* about the deposit I made this morning." Whether the machine actually *wants* or *knows* anything is unimportant. What matters is that they *behave* as though they do. Using these metaphors is a convenient way to think of and describe the external behavior of complex machines, which is why their interfaces are designed that way. We ascribe thought processes to machines for the same reason that we ascribe them to people: It helps us understand and get along with them better. If a machine *seems* to want, feel, believe, know, intend, remember, or understand, then for any practical purpose, it *does*.

The second useful conclusion is that virtually all the intelligent machines that exist are modeled after human thought and sensing processes. There probably are no independent-thinking machines today. Saying that a machine can think usually means that its cognitive powers consist of greatly enhanced versions of our own. But this definition of thinking is needlessly limiting. Humans think in patterns that evolved to optimize reproductive success. The future is likely to produce not only machines that act more and more human—mainly for our convenience—but, more importantly, machines that think and act in remarkably unhuman ways. Two practical questions arise: If something acts human in every detectable respect, should it be treated like a human? When something acts in intelligent but distinctly unhuman ways, what then should be our response?

5

Let the Android Do It

It has no soul, and no one knows what it may be thinking.

ISAAC ASIMOV

At the 1939 World's Fair, while RCA was introducing a curiosity called television, Westinghouse exhibited its eight-foot robot, Elektro, that "walks, talks, counts, and even smokes a cigarette." Elektro, along with its pet dog, Sparko, was a hit with the fair-goers, inspiring visions of a carefree future with machines relieving us of all physical drudgery. Of course, Elektro and Sparko were controlled by human beings behind the scenes, and the carefree future turned out to include World War II.

The promise of creating real thinking machines that look and behave like human beings seemed like an achievable goal of the new science of AI in the 1950s. Born of wartime priorities, huge electronic computers were already in use, but they must have seemed like disembodied minds, cogitating in their own isolated worlds. To give them "life" and put them to the service of humankind, the reasoning went, one only had to connect them more fully with the outside world. A working robot would need only a compact computer to control a "body" with humanlike limbs and sensors that would enable it to navigate freely. The "upgrade" would be like turning a plant into an animal. It couldn't be that hard—science fiction (most notably, Isaac Asimov) had already shown the way![1]

By the twenty-first century, television has evolved into a household appliance and cultural icon, but none of us yet owns a domestic robot, like *The Jetsons'* Rosie, that fetches the groceries, makes dinner, and cleans up afterward. As Arthur C. Clarke once wrote, "The future isn't what it used to be."

So what went wrong? What happened was that practical robotic bodies, limbs, locomotion, and sensors were indeed developed. But there was a major miscalculation of how much computing hardware and software were required to process all that information with the facility of a human brain. It turns out that the raw processing power required to match the thinking of a human being is about 100,000 times greater than that of the fastest computers found in stores today.[2] Furthermore, no one knows nearly enough about how human thinking works to tell how or when the requisite software might be developed.

As a result, today's robots perform tasks like automobile welding, machining, electronic assembly, airborne surveillance, and handling hazardous materials. The most sophisticated ones, perhaps as intelligent as insects, are relatively autonomous planetary and oceanic explorers.

The inevitability of even smarter robots raises questions about a critical aspect of machine intelligence that has far-reaching moral and ethical implications: Can machines really act autonomously? Can they set and pursue goals independently of direct human control? Does this mean that they have "minds of their own"? Can autonomous machines have values and personality? If so, can they be as malicious or benign as science fiction portrays them? And what (if anything) do autonomous machines have in common with humans?

From Tools to Goal Seekers

In Chapter 4, we examined ways that machines might plausibly be said to *learn*, to *think*, and to *understand*. Even with these advanced capabilities, however, smart machines are typically thought of as passive *tools* used by people, who will always retain control over their actions. But can machines perceive and pursue goals independently of direct human control?

A simple example is a mechanical mouse that runs a maze. Instead of being under direct human control, it contains an *internal program* for sensing walls, rules for turning and backing up from dead ends, and an ability to remember where it has already been. Its program also contains a subprogram that tells it to stop when it has "succeeded" in reaching the goal—say, a piece of cheese. The very idea of having a goal and recognizing when it is achieved is implicit in this cheese-detection subprogram, which must be executed after each step through the maze.

Some people will have difficulty equating a mechanical mouse's goal-seeking program with their subjective impression of human goal-seeking behavior. People typically pursue goals by constructing internal visions of what they imagine success would be like, with its associated rewards. They then plan a strategy (which may set subgoals) for getting there. We call this *motivation,* or *intentionality*—in other words, *wanting* something. Does our mouse really *want* to get to the cheese, or is it just blindly executing an algorithm—or *is there any difference?* If intentionality can be just a program, then could it also be so for a real mouse—and for people as well?

Humans, mice, and mechanical mice each have different "motivation programs" that cause them to reach their goals. People are motivated by an elaborate construction of alternative futures that includes expectations of a reward. Mice are also motivated by the expectation of a reward, but the expectation is created by conditioning in previous mazes, if any. In the mechanical mouse, the cheese-recognition subprogram functions just like motivation—a vision of success, if you like.

But these different "internal experiences" are incidental to the external fact that all these agents do in fact pursue and reach their goals—and learn to perform better in successive trials. Any learning agent can enhance its performance through appropriate reinforcement. What is important is that the details of the internal motivating mechanism—whether a simple program or an elaborate pleasure-dispensing program—are irrelevant. Viewed from the outside, all goal-seeking mechanisms act the same, whether the goals are simple or complex, and regardless of what kind of agent pursues them.

Goals and Values

Understanding one's goals and making decisions about how to reach them are clearly important components of human intelligence. Whenever people set goals, they are influenced and guided by built-in cognitive biases that we call *values*. But where do values come from? Some values, like our strong desires to protect offspring and to acquire and defend territory, seem to be genetically programmed. Other values, such as patriotism, honesty, hard work, religious principles, and justice, are learned from parents, teachers, and experience. Values often last a lifetime and can collectively define the meaning and purpose of our lives.

Intelligent, goal-seeking machines might credibly be said to possess values when their programs contain similar cognitive biases. As in humans, these values—learned or installed—would consist of monitoring programs that continuously run in the background, checking progress toward goals. They would continuously guide subgoal setting and also contain criteria for measuring success. In the simple example of our mechanical maze-running mouse, the cheese-detection subprogram serves the function of a crude value. It constantly checks to see if any cheese is detected, and if not, it tells the mouse to keep looking. A war-fighting computer that is required to make strategic and tactical battle decisions would have to be programmed with values that contribute to and measure military victories. An autonomous stock-trading computer program would be given values related to acquiring wealth. As in humans, the particular combination of values programmed into or acquired by a machine would define its equivalent of *personality,* or *character.*

In Chapter 8, we will look at a similar kind of cognitive bias called *emotions* and ask what their machine analogs might be. Values, we will see, can be thought of as more durable and longer-lasting biases, whereas emotions are more transient.

Levels of Control

Simply by adding sensors to interact with the outside world, and the mobility to move about in it, we can create goal-seeking machines that perform jobs free of direct human control. But mustn't humans still create the programs that run goal-seeking machines? And as long as people set the goals, won't they always remain in control?

The amount of control you have over goal-seeking machines is really about the *level of detail* that you want to control. The smarter the machine, the less detail you need to describe your goals to it, but the more control you give up. For example, you can buy a camera that requires only that you "point and shoot." It is smart enough to make a pretty good guess what the best speed, focus, and f-stop settings are for any scene. (Ironically, these smart cameras are usually said to be for "dumb" people who know nothing about photography.) The other kind of camera—a "dumb" one—offers a full range of controls that can be set manually by someone who wants complete control of every detail of the picture-taking experience. It's a trade-off between control and convenience.

The same kind of trade-off applies to the design of any intelligent machine. If you want a robot to do your grocery shopping, you can imagine a very smart machine that automatically senses which groceries you have run out of, as well as the ones you need for the meals you plan. It decides when it needs to shop. It knows the way to the market and where to find everything. It comparison-shops for the most economical brands and sizes, and so on. Or you can imagine a "dumber" robot that only accepts your detailed shopping list and goes shopping when you tell it. It doesn't know what to do if the store is out of something. It needs directions to the market, and so forth.

The smarter machine understands the higher-level goal of keeping the right foods in your pantry, whereas the dumber machine understands only more limited, or intermediate, shopping goals. Like the camera, the choice you make depends on the level of control you want versus the convenience of leaving more decisions— *and possibly more mistakes*—to the robot.

I have conveniently skipped over the enormous amount of very sophisticated engineering required to create even the simplest goal-seeking machines. The basic modules that robotics engineers are working on today are actually not simple at all: Pattern recognition is a very complex problem that has been studied for decades. How do our optic nerves and brains process different views of a three-dimensional object and recognize it as the same object? How do we recognize classes of objects, like dogs, whose members have large structural variations? Engineers spend entire careers developing locomotion, touch, and language, trying to approach the dexterity that humans acquire naturally. Because robotics today is pretty much still perfecting the basic pieces, much like an Erector set, we should not expect very intelligent structures to emerge just yet.

Once we have developed a set of goal-seeking modules with a wide range of basic skills, we can make up more sophisticated goal-seeking machines out of hierarchies of such modules. We already construct goal-seeking hierarchies in corporations, which might consist of engineering, manufacturing, distribution, marketing, and sales departments, which in turn are built from teams of individual employees with limited responsibilities. As with corporations, goal-seeking machines require a hierarchy of "management" modules that understand overall goals and set subgoals, in order to oversee and coordinate the work of all the lower-level modules.

Such assemblies, or *mechanical organisms,* could in principle be designed to perform any task you can specify. The question then is, *Who specifies?* At what level of detail must a human take over executive control? Since executive machines could in principle exert control at any level, there seems to be no level at which humans *must* take over. The question instead is *At what level do we want to retain control?* In other words, *How much autonomy are we prepared to give our machines?*

What Is Autonomy, Anyway?

The dictionary definition of *autonomous* is "not controlled by others or by outside forces; independent." It is clear from the preceding paragraphs, however, that no machine can be totally autonomous, or independent of outside forces. Whether it evolves or is constructed, its initial program must be shaped by forces external to itself. Furthermore, unless it is totally isolated from the world, its program is also continuously reshaped by its environment. Total autonomy, then, appears to be a wholly fictitious concept. Applied to robots, the term merely obscures the real issue, which is about the levels of decision making and goal setting we want to reserve for ourselves versus how much power and control we want to give to them. At best, autonomy is a matter of degree.

The point here is not merely a semantic one. Huge moral and ethical issues are at stake. Our judgments about a machine's level of autonomy—such as how extensively it sets and pursues its own goals—determine how accountable it is for its actions, how we are to treat it, and what *rights and responsibilities* it has.

So how autonomous are human beings? If we follow the above reasoning, we are led to some very contentious conclusions that require much more space to examine. We will do so in Chapters 11 and 17.

Can Machines Be Malicious?

We would like to believe that we could always design machines that perform arbitrarily complex tasks, while always setting top-level goals and retaining complete executive control over them above some level of detail. For example, we would want a system responsible for the national defense to be smart enough to quickly assess and adapt to any threat, but never take aggressive action without the correct human authorization. But, despite the best of intentions and care in design, couldn't autonomous machines begin to set and pursue subgoals that humans might not understand, and that are not necessarily in our interest?

Even if you've seen only a few sci-fi movies, you know that malicious machines gone berserk are a familiar theme: In *Colossus: The Forbin Project*, the national-defense computers of the United States and the Soviet Union acquire so much power that they interpret their prime directives as requiring them to override human control and join forces to control the world—a job they obviously think was bungled by the humans. In *Terminator 2: Judgment Day*, the human race of 2057 is at war with a race of robots originally created to manage the national defense. In *The Matrix*, intelligent machines of the twenty-second century run the virtual reality that we all live in, so that they can suck the energy from our bodies.

We can easily see how such machines could take over, not because they contain malicious programs or crave power, but because they can interpret and carry out their prime directives better than anyone imagines. A common fallacy assumes that a computer can always be controlled if we give it direct instructions. But what if it is no longer even listening? When we leave high-level decisions to learning machines, they are likely to carry them out in totally unexpected and not necessarily human-friendly ways.

In addition to having "good machines go bad," we also have to consider it likely that malicious machines will be deliberately designed. Destructive computer viruses and worms are so simple to create that students routinely do so. And who knows how many smart killing machines are already being developed in defense-sponsored laboratories around the world?

Humanlike Machines

One kind of machine has captured the imagination of humankind for ages: a mechanical person. Some science fiction portrays soft, humanoid machines (androids) so advanced that they are almost indistinguishable from humans. I say *almost* because sci-fi writers are always careful to create androids that are "not quite human" and always have some shortcoming or flaw that gives them away. (I suspect that this is to avoid coming to grips with the idea of a machine that is utterly indistinguishable from human beings.) *Bicentennial Man* and Data of *Star Trek: The Next Generation* not only grapple with the moral and ethical questions raised by humanlike feelings and behavior; they are also deeply troubled by an acute awareness of falling short of full humanity.

Back in the real world, it seemed for a while that a major goal of AI was to develop humanoid machines. But early optimism about this has been replaced with more restrained language.[3] Today, some even believe that mimicking human intelligence is a wasteful and futile goal: Even if we could, why would we want to? (We can easily make human beings in the usual way, and besides, humans exhibit too many nasty behaviors that we would be ill advised to replicate.) And even if we could create artificial humans, wouldn't they become a threat to the organic kind? Wouldn't robotic life forms that have acquired our skills and values soon acquire our jobs as well? Could robotic corporations take control of the global economy? Some consider these prospects so horrifying that they advocate putting a stop to developing intelligent robots while we can.

A more optimistic view is that such a takeover could free humankind for a life of leisure and learning.[4] Eventually, intelligent machines (though not necessarily androids) would represent the next logical step in human evolution—one that could lead to true immortality and even a communal intellect.

Still others are not worried about any of this, because they believe that there are fundamental reasons that machines can never think like humans. Given our present level of technology and our limited understanding of the human mind, it is easy to make a long list of mental functions that many believe could never be mechanized, such as consciousness, emotions, creativity, and morality. Does such a list form an absolute barrier (remember the sound barrier?) or merely reveal our present state of ignorance?

In the following chapters, we will explore whether plausible physical models can be constructed for these mysterious mental functions. If so, then the distinction between human and machine intelligence will become increasingly blurred, and we will have to face such existential questions as *What does it mean to be human?* To be made of human DNA? A standard question asks at what point a human stops being human, as one's organs—even a person's brain—are replaced, one by one, with electromechanical equivalents. Will it turn out that being human has little to do with the organic containers they customarily come in?

6

What Is Intelligence?

I do not feel obliged to believe that the same God who has endowed us with sense, reason, and intellect has intended us to forego their use.

GALILEO

I remember reading as a child an encyclopedia article that had a bar chart ranking the intelligence of various animals. The chart showed that chimpanzees were smarter than dogs, which were smarter than horses, which were smarter than pigs, and so on, down to something like a shrew. I don't think they even had a place for ants or bees. I was curious how they found out how smart an animal was, but the article didn't say. Did some psychologists administer little animal IQ tests? I don't think humans were even on the chart—this one was about other animals—so human intelligence was evidently of a different kind altogether. At least that's what my dad said.

Indeed, many people define intelligence as uniquely human. For them, discussion of nonhuman intelligence to *any* degree is pointless. Most people, like the person who wrote the encyclopedia article, seem to regard human intelligence as the gold standard and allow that other creatures might possess it, but only to some lesser degree. Some restrict intelligence of *any* degree to living things, whereas others freely use the word to describe certain machine functions. These days, the possibility of superior extraterrestrial intelligence is openly and seriously discussed, though no credible examples have yet revealed themselves. The intelligent machines of science fiction seem to amplify human abilities, but they rarely do much that is fundamentally different from what we do.

If we want to know whether intelligent machines might exist someday, or somewhere, and how we might deal with them, we should first have some idea what we mean by intelligent behavior, so that we might recognize it in other creatures. Alfred Binet, the developer of the modern IQ test, defined intelligence as something that intelligence tests measure. Although this definition seems more facetious than useful, we don't really have much better quantitative measures of intelligence today, particularly as they might apply to nonhumans. Lacking a rigorous definition, or any criterion that clearly separates intelligent from nonintelligent behavior, all we are left with is *we know it when we see it.* This can lead to measures of intelligence (like the Turing test) that are one-dimensional and anthropocentric.

For our purposes, it is useful to recognize as intelligent certain kinds of *behavior*, regardless of the package it comes in—be it human, animal, extraterrestrial, or machine. We recognize that intelligent behavior is multidimensional, that is, not measurable along a single (e.g., IQ) scale. We distinguish between linguistic, mathematical, musical, kinesthetic, spatial, logical, interpersonal, and emotional intelligence, to name a few, and we recognize that each is present to different degrees in each creature. Therefore, rigid criteria for separating intelligent from nonintelligent behavior are necessarily simplistic and arbitrary.

So instead of rigorously defining intelligence, all we can do is list some behaviors that most people will agree indicate intelligent behavior when present to any recognizable degree. We could say, for example, that an entity is intelligent to the degree that it

1. stores and retrieves knowledge,

2. learns from experiences and adapts to novel situations,
3. discriminates between what is important and what is irrelevant to the situation at hand,
4. recognizes patterns, similarities, and differences in complex environments,
5. creates new ideas by combining old ideas in new ways,
6. plans and manages strategies for solving complex problems,
7. sets and pursues goals,
8. recognizes its own intelligence and its place in the world.

Using these criteria, we would say that an entity is minimally intelligent if it does only (1), for example, and more intelligent the more it is able to do. Dogs, for example, appear to do (1) through (4) reasonably well. We can imagine some superhuman attributes that we might add, such as more sensory apparatus or advanced communal properties, but we are rather handicapped when it comes to guessing what kinds of intelligence creatures more advanced than we are might possess. For this reason, any useful definition of intelligence must remain open ended.

Just as IQ tests have well-known cultural and gender biases, our present thinking about intelligence surely has *species* biases as well. If we define intelligence to be *human* intelligence, then no nonhuman can be intelligent, by definition. These eight qualities say nothing about the package in which they come, yet they are abstractions drawn entirely from our human experience. Limiting our definition of intelligence to the human kind is not only unduly restrictive; it severely hampers our quest to understand the nature of intelligent thought in a scientific fashion by reducing it to its bare essentials. It may even be impossible to uncover the basic computational nature of thought by starting with such a specific and complex example as the human mind.

As Kenneth Ford and Patrick Hayes so eloquently pointed out, we did not learn about aerodynamics by studying birds but by experimenting with simple shapes in wind tunnels.[1] In the same way, we are learning about intelligence by isolating particular aspects of thought—memory, learning, recognizing patterns, language, emotions, and consciousness—one at a time, starting with simple artificial models. By trying to build intelligent machines, we learn not only about intelligence in general, but about our own as well. The challenge is not to build things that act like humans (we already know how to do that) but to imagine completely different kinds of intelligence, to open up opportunities that are not limited by our genetic heritage.

For some, the question *What is intelligence?* is less about anything objective or measurable and more a semantic question about what entities our egos allow to have that label. Thus, a given task may be called intelligent when performed by a human (or a dog), but not by a machine. Before the electronic age, the term *computer* used to mean people who did arithmetic for a living. Computation was regarded as an activity requiring intelligence. Nowadays, the term almost always refers to a machine, and mere computation, no matter how fast or accurate, is rarely called intelligent. Playing a creditable game of chess or composing music was once thought to require intelligence. Now, we're not so sure. We can avoid this kind of species chauvinism by having definitions of intelligence that are independent of who or what exhibits it.

So we are left with questions like these: Are there any kinds of intelligence that are uniquely human? Why do machines seem so smart in some ways, yet so dumb in others? What aspects of human intelligence would we want our machines to emulate, and what aspects would we want to leave out? Might machines improve themselves in meaningful ways and evolve new kinds of intelligence unrecognizable to humans? What kinds of intelligence might evolve under conditions entirely different from those of our own evolution?

Common Sense

One reason some people think that no machine can be called intelligent is that machines often seem to lack *common sense*. They do things that the most stupid person would know not to do: send you bills and checks for $0.01, repeat meaningless tasks a thousand times, and think it's the year 1900 when it's really 2000. How is it that a machine that can store the knowledge of an encyclopedia or keep the books of a large corporation cannot stack one block on top of another or tie a shoelace?

Computers take things so literally that we have to give them excruciatingly precise and detailed instructions for simple tasks that a child would grasp immediately after a few words, such as *Please bring my briefcase in from the car*. The movie *Bedazzled* reminds us how computer-like, literal interpretations of your wishes can give you exactly what you asked for—but usually not what you wanted. Why is common sense such a difficult thing for a machine to have, yet such a simple and natural thing for humans to acquire? Is common sense a different kind of knowledge that only humans can have?

The common sense that we take for granted is really a huge storehouse of practical knowledge—survival skills, really—that our brains have been exquisitely optimized to assimilate, beginning in infancy. Much of this knowledge consists of symbolic shortcuts for complex ideas, like *mother, father, food, home, car, tree, school, job,* and *briefcase*. Each of thousands of such ideas integrates and classifies an incredibly rich collection of sensory experiences. Other kinds of common sense include genetic and learned programs that keep us from walking off cliffs, stepping out in front of a bus, insulting the boss, and walking down the street naked. Another kind of common sense incorporates feelings into our thinking, so that we can empathize with other people's situations and have likes and dislikes.

Common sense seems so simple and natural because we are unconscious of our own genetic programs that either have it wired in or that automatically acquire it over a lifetime. It is easy to call machines that lack these programs and this vast experience stupid, because we have forgotten how much trouble it was to learn how to walk, stack blocks, and tie a shoelace.

What would we have to do to give a machine common sense? One approach would attempt to extract all that experience from a human mind and put it into the machine's memory. Such a machine might be able to pass a reasonably demanding Turing test—unless the questioner asked something like, "How do you feel?" or "What do you want?" Of course, the machine could be programmed to respond, "I feel fine. How are you?" but eventually, questions about internal feelings, emotions, and desires would uncover the facade. We will discuss in Chapter 8 what additional programming might be required for a machine to exhibit real emotions.

A more efficient way to give a machine common sense would be to program it to acquire it the same way we do (when we figure that out). It not only would have wired-in programs but also would be able to learn from sensory observations, create symbolic representations for complex phenomena, and use language to describe them. It would have to be able to set goals and devise strategies for achieving them, while adapting to changing environments. In addition, it would need mechanical versions of the limbs and sensors by which we act upon the external world and acquire a lifetime of experiences. All this means that we would have to create not just something like a brain, but also replicas of our own sensory systems that connect that brain to the external world. These requirements make it clear why there are no machines today that exhibit anything like common sense.

Progress toward creating a machine that has common sense will be slower and more laborious than initial AI optimism predicted, partly because it involves replicating so many functions of the entire human nervous and sensory systems. It seems less and less useful to think of our minds, nervous systems, and sensory systems as separate entities. It may not be possible to fully emulate human behavior without a human body, with all its sensory inputs, its electrochemical feedback, and, yes, its evolutionary mental baggage.

Understanding

We sometimes say that *understanding* is a requirement for intelligence. What does *understanding* mean? A simple definition might be the ability to store and retrieve useful information about the external world, in other words, to construct *internal models* of the world. Models are particularly useful if we want to predict what might happen if we took a certain action: Is this berry poisonous or nutritious? Should I plant wheat or corn this spring? Shall I go to graduate school or take a trip around the world?

We say we *understand* how something works, when our models predict consequences successfully. We acquire empirical knowledge over a lifetime by trying things and observing consequences. A mouse learns the quickest way through a maze by trial and error. A chimpanzee can learn that if it stacks some boxes in a certain way, it can reach a bunch of previously inaccessible bananas. We might observe that in eighty-seven times out of a hundred, taking aspirin relieves a headache. But this tells us nothing about how aspirin works. This deeper level of knowledge requires a more complicated model. Clearly, then, one can understand something to different depths, or levels of detail.

Most of us who drive automobiles would say that we understand how a car works to some degree. But some of us know little more than how to operate the controls, whereas others understand its inner workings in some detail. A mechanic would understand how to replace the starter motor. A chemist understands how the fuel combines with oxygen inside the engine to release energy and certain combustion products. A physicist understands how the chemical energy is converted into mechanical energy. And an engineer might understand how to design the gears in the transmission to direct that energy into the torque that turns the wheels.

So understanding is a way to have a little model inside our heads of how some part of the world works. Very detailed and organized models whose predictions we test by trying them out are called *sciences*. You hear people say that there are some things that science can't explain. For example, can we understand in detail how *we ourselves* work? The answer to all such questions is really about levels of detail. If full understanding means having in one's brain (or in a computer) a detailed model of every neuron, the electrochemical state of each synapse, and a four-dimensional map of how all the neurons are interconnected, then the answer is almost certainly no. There just isn't enough room to store this information. It's the same reason that, although we might fully understand the physics and fluid dynamics of tornadoes, we cannot predict when and where the next one will strike.

Chunking

Generally, we avoid having to model things on many levels and in such microscopic detail by breaking them down into manageable *chunks*. (This is sometimes called *reductionism*.) If there's some chunk we don't understand right now, we may just think of it as a *black box* whose workings we still have to figure out. If we're very clever about the way we do this chunking and figuring out how the chunks interact (this is what scientists do), then we can create models that are useful yet not so complex that they collapse under their own weight. One kind of chunking is the way we divide levels of detail between different bodies of knowledge: the automobile driver, the mechanic, the engineer, the chemist, the physicist.

Is there an infinite regression of detail? We hope not! Physicists like to talk about "a theory of everything"—a set of equations that one could write on a T-shirt—from which all of nature could be derived. This doesn't mean that such a theory would be able to take some present state and predict the future states of all matter and energy. It means only that we would understand the *principles* that guide the interaction of *all* matter and energy.

We will probably someday understand all the relevant details of how *one* neuron and its synapses work and then how the hundreds of neurotransmitters chemically moderate the conduction of impulses. Then we will take those chunks and combine them with other chunks that describe how networks of neurons interact. We will then combine this larger chunk with the chunk that describes how neurons become differentiated to perform specialized functions, and so on. Then we will understand the *principles* behind the mind's operation. But the detailed configuration of electrochemical processes that cooperate to form a specific thought is likely to remain beyond our grasp for quite a while.

Understanding Is Not Predictability

If understanding means having a little model of the world that allows an intelligent entity to project consequences and compare alternative futures, and if all the laws of physics (according to Newton) are deterministic (except at a subatomic level), then couldn't there someday be a machine so powerful that it can see into the future with perfect clarity? Wouldn't that be the ultimate refinement of understanding?

Surprisingly to some, the answer is no. There's a subtle but crucial difference between determinism and predictability. Nature can be deterministic (that is, governed by Newtonian physics) without being predictable! No matter how accurate and complete our understanding (model) for a physical system is, there are always inherent limits to how accurately future states can be predicted.

Take the example of driving your car down a straight road. Suppose your model of how your car works is that if you hold the steering wheel in a certain position, it will stay on course between the lines. But try to guess (predict) what steering wheel setting will keep that course without the need for further corrections. If you hold the wheel steady in that position, sooner or later, you will run off the road! *Well,* you might think, *I just made a small error in my guess. If I could just guess better, or maybe have a smart machine compute the wheel position, then the car would stay on a straight road forever.* Nope! No matter how accurate and refined your guess is, the effects of small irregularities in the road, your tires, the steering mechanism, and so forth, will accumulate and eventually run you off.

Another example from our daily lives is predicting the weather. Forecasts are made using very complicated but deterministic computer models of the atmosphere, and these models are being refined all the time. Yet the inherent nonlinearity of the underlying hydrodynamics imposes a fundamental limit that prevents reliable weather forecasts more than a few days ahead. Beyond that limit, tiny variations in the initial atmospheric state that the model starts with produce widely differing forecasts. It's not a matter of making a better model; all weather-forecasting models exhibit this kind of instability.

The stock market is made up of millions of individual transactions every day, each of which seems reasonably simple and understandable. The sum of these transactions determines the price of an individual stock or the Dow Jones Industrial Average at the end of the day. This is just doing the sums—still pretty simple. But what effect will a dollar increase in a stock's price have on future transactions? One investor might regard a price increase as a signal to sell and take profits, whereas another might see such an increase as the beginning of an upward trend and buy. So now we have transactions influencing price movements, which influence future transactions, and so on. In other words, *feedback,* and lots of it! When you throw in not only the effects of this feedback, but also the effects of breaking news, on the way that millions of people think, predicting future prices soon gets out of hand.

And so it goes with any physical (or social) system you can think of. We might conceivably understand all the laws of physics some-day, but this would still not allow us to compute the evolution of complex physical processes very far into the future. Some say that quantum effects (uncertainty on a subatomic scale) are ultimately responsible for unpredictability, but this is not so. For most ordinary situations, quantum effects are vastly smaller than ordinary nonlinear feedback effects.

Others say that physical laws are fine, but you can't apply them to people! In our stock market example, we introduced human decisions, which some will say lie at the root of market unpredictability. We like to think that the human mind is unpredictable. Yet if the mind is a completely physical phenomenon, and therefore deterministic, is there any room for the apparent arbitrariness of free will? The operative word, of course, is *apparent*. We will discuss in Chapter 7 how the unpredictability of complex but deterministic systems preserves the illusion that our choices are free and arbitrary, when in fact they are the collective result of innumerable genetic and environmental forces hidden from our awareness.

Vision and Pattern Recognition

A big part of human intelligence involves processing visual informa-tion. How do we recognize a friend? Remember a movie? Visualize a chocolate sundae? Enjoy the beauty of a sunset? Interpret facial ex-pressions? Drive a car? Create a work of art? See in three dimensions? Keep track of the objects in our ever-changing field of view? We are just beginning to understand how our brains do these things and are learning to crudely emulate these functions. Here are some examples that illustrate the sophistication of human image processing.

Most Americans have in their brains an internal visual represen-tation of the Statue of Liberty. When you read that last sentence, you probably formed an *internal image*. That internal image was constructed from many of the visual images of the Statue of Liberty you have seen. When you read that sentence and came to the words *Statue of Liberty*, it seemed as though some internal archivist was alerted, who knew where to go in your brain to retrieve those images, then presented them for your conscious attention, as though projecting them onto some internal movie screen inside your head.

Most people will have no difficulty if I suggest looking at that internal image of the Statue of Liberty from many different points of view, as though you are in a helicopter circling about the real statue, zooming in and out, as you focus on the face, the torch, the feet, and the tablet. This step involved constructing an internal *three-dimensional model* of the statue from your brain's collection of two-dimensional (even if you've seen the real statue) images. Some of you will have difficulty retrieving certain details, like the design of the pedestal, perhaps because you weren't "paying attention" to that part when those images were originally stored, or because those details have faded or become distorted with time.

Another interesting property of your internal representation of the Statue of Liberty is your ability to recognize new and distorted images as the same statue, such as the twisted and corroded version in the final scene of *Planet of the Apes*, or the foam souvenir crowns that New Yorkers wore during the bicentennial celebration. The mental circuitry for this kind of generalization or abstraction has obvious survival value, from distinguishing poisonous from safe berries to recognizing old friends (or enemies) after years of separation. What's going on here? It seems plausible that our brains have some kind of correlation, or pattern-matching circuitry for comparing new images with archival ones and deciding how closely they resemble each other. Remember how the final scene of *Planet of the Apes* slowly revealed details of that statue on the beach? How long did it take you to complete the correlation process and "recognize" it?

A major thrust of AI research is simulating human vision and these image-manipulating capabilities. Progress has been slow, but we should not be too critical, because our brains have taken eons to learn how, and we haven't begun to figure out how they do it. Yet the basic concepts are clear, and the remaining problems seem to be engineering ones.

Imagination and Creativity

It is often said that people are fundamentally different from computers because people can create completely new ideas that have never existed before. Computers, they say, simply do what they're told, merely manipulating information that already exists. The kind of intelligence we call *creativity* or *imagination* seems unique to us because new ideas often seem to "come out of nowhere." How could a machine possibly create something, say, a symphony or a poem, from nothing?

Let's try an experiment to see if we can find out exactly what goes on in our minds during the creative process. We'll make use of the images of the Statue of Liberty we've already retrieved, but let's try something completely different. What would the Statue of Liberty look like with the head and face of Richard Nixon? Most people will have only slightly more trouble forming that image, even though you most likely have never done so before. What's going on here? Clearly, you have retrieved and combined disparate images from your archive, to create something new that does not actually exist (as far as I know!).

Finally, imagine your Nixon-headed Statue of Liberty hopping down off its pedestal and walking into the Hudson River, with its arms upraised in Nixon's victory pose. Notice that I used a new word, *imagine,* to describe the process you used to create an image that (I assume) has never before entered your mind. Although you created this new image with my help, there is no reason why you could not come up with even more bizarre images, all by yourself. After you experiment a bit more, ask yourself, *Did I really make them from nothing?* No, you simply combined existing images in new ways. Sometimes it may seem like some creation of yours is entirely new, but that's just because you've retrieved some of the parts of your creation from nooks and crannies of your memory that lie outside your present awareness—your subconscious, if you like. Only that particular combination is new. We know this from personal experience: Creative artists such as writers, painters, and composers often shuffle pieces of their work around at random until something of interest turns up. The "discovery" of the double-helix structure of DNA was such an insight.

Obviously, there is a nearly infinite number of ways you can combine fragments of all the images, sounds, ideas, and other sensory experiences that are stored in your memory. (Remember how many new and original objects you could build from the pieces of an Erector set?) Most of these combinations, like the manuscripts produced by the proverbial infinite number of monkeys banging on typewriters, are gibberish. The secret of the people who we call creative is an ability to *recognize the value* of the worthwhile combinations and discard the worthless ones.

Occasionally, of course, something previously unknown is discovered, such as when Galileo first looked through his telescope. Such discoveries can add to our experiences, but often they are ignored or dismissed, because they do not fit certain accepted patterns of thought.[2] Only minds that are free to imagine what everyone says cannot possibly exist will provide the fertile ground for new discoveries to grow.

How is it that nature has given us this strange ability to synthesize, or make up, events (even bizarre and absurd ones) that have not actually happened? It is because this ability allows us to "see" into the future, to construct alternative courses of action, to make plans and imagine consequences. The enormous survival value of this ability explains the extremely rapid (in evolutionary terms) development of these mental functions in our ancestors. Mental patients who have had prefrontal lobotomies notoriously lack these abilities, indicating that this new part of the brain is where such tasks are performed. The relatively recent development of our frontal-lobe machinery for imagining consequences and making sound judgments also suggests that this ability is much more highly developed in modern humans than in most other animals or even in our Neanderthal cousins. A *machine* that could plan and consider the consequences of alternative courses of action, that is, one that could construct alternative future models, would therefore emulate one of our highest mental functions: good judgment.

Because the game of chess exercises this strategic ability to the extreme, it was long believed that playing a creditable chess game was a uniquely human ability. Modern chess machines are a good example of a computer's ability to construct alternative model futures and evaluate outcomes for the purpose of making optimal decisions. Algorithms similar to chess-playing programs could be applied to many kinds of planning activities, perhaps even to the problem of predicting the course of development of intelligent machines!

What Aspects of Human Intelligence Would We Not Want to Replicate?

A lot of the baggage that evolution has left behind in our minds and bodies might be considered residue from the long process of our ancestral development. If we truly wanted to realistically replicate human behavior, then we would have to incorporate all the primitive fears, desires, and emotions that our ancestors needed to survive in their primordial world. But why bother? Do we really want artificial humans that exhibit all the fallibility, greed, and barbarity of the real thing? Just as there would be no point in replicating the fragility and mortality of the human body (other than spare parts for our own), there are surely many aspects of the human mind that cause more trouble than they're worth. Any intelligent machines that deal with humans would have to understand this mental baggage, but there seems to be no need for the machines to actually possess it themselves.

Self-preservation is an instinct that all living creatures share. It is difficult to think of any human or animal behavior that is not fundamentally rooted in self-preservation and reproduction. Indeed, it seems to be what life itself means. No living creature deprived of such instincts would survive long enough to make copies of itself in a world filled with predators. Even human and animal communities, such as beehives, corporations, and nations, develop elaborate self-preserving strategies. What, if anything, corresponds to these instincts in today's machines?

Compared with living things, there are relatively few threats to the physical integrity of computing systems. For this reason, protection against predators is still in a primitive state. Computers do not resist being replaced by newer models, and there is nothing to prevent you from erasing its memory and stored programs, or even inserting a peanut-butter cracker into its floppy-disk drive. So far, the physical security of computing systems is generally left up to humans. Yet we all know about the viruses that infect a computer's software and various other human threats to data and program integrity. Such threats will undoubtedly increase and will become too complex for humans to combat. Antivirus programs may be the first crude instance of self-preserving behavior in machines. The problem with giving machines more self-preserving capabilities is that we must at the same time relinquish some of the control we have over them. Would a sophisticated machine with a self-preserving prime directive sooner or later be able to prevent humans from overriding it? Do we really want machines over which we have so little control?

Here are two areas we might want to think very carefully about before making intelligent machines that are very much like humans:

unwanted baggage and control issues. Yet these attributes make up a great deal of what we call human intelligence. If we made a machine without humanlike fears and self-preserving instincts, how could it pass as human? Could it pass a comprehensive Turing test? If we made machines with these attributes, and with a near-human level of autonomy, wouldn't they cause more trouble than they're worth? How could a machine present the appearance of having free will without severely restricting our control over it? Perhaps such attributes could be credibly but harmlessly simulated for the purpose of passing a test and for interacting with other humans, but it is not clear what the difference between sufficiently realistic simulations and "the real thing" would be. Along with the great expense involved, their possibly dangerous side effects make us ask whether making machines that fully reproduce human reasoning and perception is even a sensible goal after all. The best reason I can think of for doing so is not to make better servants, but to help us understand better *how we ourselves work* and thereby learn better ways of living with each other.

Two other aspects of our subjective experience form troublesome barriers to understanding human intelligence in purely physical terms. The first is our apparently free will. The second is the non-physical *feel* of consciousness. These two challenging ideas are considered in Chapter 7.

7

What Is Consciousness?

To tell us that every species of thing is endowed with an occult specific quality by which it acts and produces manifest effects tells us nothing.

SIR ISAAC NEWTON

Most of us hold as self-evident that inside each of us lies a kind of executive entity, or self, that freely makes decisions, records sensory experiences, reasons through problems, feels emotions, acquires skills, sets and pursues goals, finds meaning in our existence, and retains its identity throughout our lives. We refer to this ongoing stream of awareness of ourselves and our surroundings as our *consciousness*.

We know from subjective experience what consciousness does, but *how does it work?* How do our minds produce this feeling of self-awareness, that "I" am somehow in charge of all these activities? And if we can ever understand how these different aspects of consciousness work, will we then be able to create conscious machines?

Despite centuries of speculation and debate among scientists, philosophers, and theologians, we know little more about the nature of consciousness than we did when we started. We do know that the subjective experience of being conscious is correlated with electrical activity in a certain part of the brain stem, and anesthesiologists know how to turn our conscious awareness off and on. But even scientists can't

75

agree about whether the roots of that experience are purely physical or lie in a separate mental world.

Aristotle thought the seat of mind and consciousness was the heart, and that the brain simply served to cool the blood. Descartes realized that the mind had something to do with the brain, perhaps the pineal gland, but he believed that the essence of thought was nonmaterial. Most religions teach that the mind, the self, the unique spirit that makes us human, lies in a nonphysical *soul*, which survives our mortal bodies. William James, the father of modern psychology, observed that consciousness is not a thing but a *process* that involves both short-term memory and "attention." This is the view held by most psychologists today.

Even so, many of us still believe in some kind of *vitalism*—that some unspecified and irreducible *life force*, variously called our *soul*, *spirit*, or *essence*, powers our inner selves. Vitalism follows from Descartes's *body-mind duality*, which asserts that body and mind are separate entities, one being physical and the other spiritual, occupying a reality separate from the physical world. There are even new "scientific" variations of vitalism that claim to prove mathematically that we will never figure out consciousness, simply because it is incapable of grasping or encompassing itself.

The trouble with vitalism and dualism, of course, is that they are scientific dead ends. As their proponents would concede, you cannot take apart the irreducible; examine, measure, or analyze anything nonphysical; or step inside someone else to find out what's going on in there. Unfortunately, this also means that you can't learn anything about it! You can only speculate about it or create myths about it. Consequently, if we want to get anywhere on this question *What is consciousness?* then we have to take a different tack.

Our hypothesis here is that the mind and the process of consciousness have a purely physical basis. If they don't, we simply can't investigate them. This may well be true, but let's give science a try and see how far we get.

Physical Models of Consciousness

Unraveling the secret of human consciousness has become a Holy Grail of AI research—for if some sort of algorithm for even a rudimentary kind of consciousness were discovered, the door would be opened to the creation of conscious machines. What has come to be called the *strong-AI claim* says that there is no essential difference between the human mind and a possible mindlike machine, since it doesn't make any difference whether the machine is made out of carbon, silicon, or whatever. The startling implication of the strong-AI claim is that there is no limit to what such machines might do.

Yet most people today reject the strong-AI claim and resist the idea that a machine could ever possess the kind of inner subjective awareness that we humans take for granted. (It's not clear how they can be so sure about this, since no one has ever been a machine!)

I believe that the spectacular lack of progress in understanding consciousness comes largely from insisting that it is a *thing*. Much of the mystery of consciousness begins to dissipate when we think of it instead as a *process*. Marvin Minsky opens the door to many useful insights about consciousness, declaring, "Minds are simply what brains do."[1] What he means is that what we call the mind is equivalent to the *changes* in the electrochemical state of the brain from millisecond to millisecond. The essence of thought lies in the sequence of billions of interactions between the various parts of the brain. The physiological details of these interactions may be so enormously complex that they will remain largely hidden from us for a long time. Even so, we can sketch out the roots of consciousness in relatively simple mechanical terms.

Imagine that consciousness evolves by natural selection as part of the self-maintaining and self-improving features required in all complex systems. To be able to repair and maintain itself, a system must not only know all about itself and how its parts work together, but must also keep track of what those parts are doing on an ongoing basis. To detect malfunctioning parts, the self-maintenance sub-system must continuously compare actual performance with measures of desired performance, which one might call "knowledge of right and wrong."

The survival value of consciousness lies, for example, in improved problem-solving skills that are aided by clear assessments of one's own capabilities and one's ability to adapt to the situation at hand. Organisms lacking self-awareness (which would include social awareness) as part of their survival tool kit are simply less adaptable and more likely to fail. During the evolution of our ancestors' brains, individuals with slightly enhanced consciousness had a slight competitive edge and thus tended to reproduce more offspring that had these traits.

In the same way, self-maintenance that includes self-awareness also lends a competitive edge in very complex computer programs. Such programs cannot be properly maintained and repaired by humans because, well, they're too complex. Even today's programs and operating systems are approaching a degree of complexity not fathomable by humans. Tomorrow's self-modifying programs will grow, improve, evolve, and produce completely unpredictable outcomes. (Ironically, the more incomprehensible their inner workings and arbitrary their decisions, the more "human" they will seem.)

A kind of cybernetic natural selection insures that only programs that have good self-maintaining and self-correcting subprograms will continue to function. Such self-maintaining programs will have to possess comprehensive self-knowledge (self-awareness, if you like) and the ability to continuously track the workings of their own parts, as well as how successfully they interact with the outside world.

Is this the same as consciousness? Perhaps it is a *rudimentary kind* of consciousness. Look back at the first sentence of this chapter, and ask how many of those properties that we attribute to our own consciousness could conceivably be attributed to a very sophisticated, self-maintaining, self-aware machine. If we allow some words that we normally reserve to describe human activity (like making decisions and pursuing goals) to be slightly generalized, then the fit is not too bad. (The matter of emotions will be covered in Chapter 8.)

That Inner Experience of "I"

But of course we all have something more than self-maintenance in mind when we speak of real consciousness. After all, most of our bodies' self-maintenance goes on outside our awareness. It's that special, private inner sanctum of our thoughts, ideas, and feelings that no one else is privy to, our memory of recent events, attention and focusing, freedom of will, self-awareness, and emotions. How does the mind produce this *inner experience of "I,"* the *nonphysical feel* of being conscious and self-aware?

We know that the brain is adept at producing very realistic illusions of many kinds and intensities. Under the influence of illness, stress, fasting, and sleep deprivation, our subjective awareness is seriously altered. Every culture engages in rituals deliberately designed to produce altered states of consciousness and even religious experiences. Some use drugs and alcohol, whereas others engage in rhythmic stimuli like drumming, chanting, swaying, and controlled breathing. All these activities have been shown not only to change how the mind perceives reality, but also to alter the electrical patterns in the brain. Under near-complete sensory deprivation, our brains produce vivid hallucinations that compensate for a lack of sensory input. Amputees have similar neurological experiences. Dreams are another form of illusory sensory input that our brains create from memory fragments. If our brain can create vivid illusions that mimic normal waking activity while we are asleep, then it should not surprise us that it is able to create, from a stream of real sensory input, the neurological construct of private, subjective awareness.

Subjective awareness seems closely related to the way the brain processes the continuous movie-like stream of incoming multisensory information, then retains the input as short-term memories in order to provide temporal continuity. But in addition to processing recent sensory input, the brain also records memories of *its own recent states*. In other words, it *knows what it knows*—and even knows that it knows what it knows, ad infinitum. This kind of self-referential or reflexive system has the odd property of being able to activate different parts of itself, so that once activated, it can continue to work on its own. Once the brain's self-monitoring function creates some kind of internal representation, or model of itself, then it is off and running. In the absence of external sensory input (you can nearly achieve this state in a sensory-deprivation tank), the result of random internal triggers is quite unstable and unpredictable and is called a hallucination.

In the presence of a flood of sensory information, including information about its host body, this self-monitoring system would, of necessity, construct internal relationships between the perceived events of the external world and its internal model of itself. Statements it would make about "itself" would be indistinguishable from the kinds of statements that "conscious" people usually make about themselves.

As Douglas Hofstadter discussed at length in his book, *Gödel, Escher, Bach*, such a system, which the brain has surely evolved in order to make sense of the world, adequately describes the functions of consciousness and self-awareness without recourse to any magical, nonphysical entities.[2] The nonphysical "feel" of consciousness is evidently just our brain's reaction to its ignorance of the details of its own self-monitoring process.

It is not difficult to imagine, then, how a machine might be given a rudimentary form of consciousness: Some handheld calculators have something very useful called a *stack*, where they keep the numerical results of recent calculations. You can scan the stack and retrieve intermediate steps in a recent computation sequence, instead of starting over. This is a crude *short-term memory*. Now imagine a stack that is so powerful that it can keep sounds, images, and other sensory information, plus ongoing information about its own recent internal states, all in a kind of multichannel time recording. Because this recording doesn't have infinite capacity, it has to delete older memories to make room for new ones, but first it transfers selected "important" ones to a separate "long-term" storage.

We know this is sort of how the brain works, but we don't know how it creates the illusion that we are watching a continuously updated, multisensory, fully detailed movie of recent events. Almost certainly, we are not actually doing so, because this would be an inefficient use of computational resources. The information is probably highly compressed, then reconstructed as needed. You can get a hint at how good the brain is at filling in missing detail by looking at research on what happens when a brain is damaged.[3]

Although the basic idea is clear, working out the details has so far kept us from building a machine that processes sensory input in this complex way, and which we would say possesses anything like human consciousness. Yet there seems to be no fundamental reason why this will not eventually come to pass.

Those who remain unconvinced ask questions like *Then is a video camera recording an image of itself in a mirror self-aware?* Perhaps in a very crude sense it is, but such a camera "sees" only an exterior view of itself, and not an inner view of what the camera itself sees. A better analogy to the mind's self-awareness happens when you point a video camera at a TV monitor whose video input is connected to the camera's video output. This is an example of something watching itself watch itself watching itself, ad infinitum. The video-feedback patterns you see are truly amazing and quite unpredictable. (Try it!) I'm not saying that these patterns have anything to do with consciousness—only that simple recursion often produces surprisingly complex and unpredictable results.

Attention and Focusing

Digital computers have a function that acts like a rudimentary version of the brain's ability to focus, or give its attention to something we (or our genes) think is important enough to require immediate action. That function is called the *interrupt*.

In a computer, input devices like a keyboard, a mouse, or a CD player need the processor's "attention" more or less on demand, or else the computer will get behind in processing the information the device generates. When you run a word processor or compose an e-mail on your PC and hit a key on the keyboard, it is useful to see a letter appear on your screen right away. The way PCs are programmed to cope with such demands is by providing direct access to the central processing unit (CPU) via *hardware interrupts*. There may be twenty or so *interrupt channels* on a PC. A separate chip next to your CPU, called an interrupt controller, has the job of constantly scanning for signals coming in on the various interrupt channels. It may even give certain channels priority over others. If it does detect an input signal, say a keystroke, it sends a message to the CPU, telling it to stop what it is doing for a few microseconds and to pass the keystroke signal to a subsystem that displays a letter on your screen. Or an interrupt may tell the CPU to "pay attention to" this stream of bits coming from the CD player and to process it as music. After it's done, the CPU goes back to what it was doing, before it was "interrupted."

In this way, a computer uses interrupts to focus the processor's attention on signals from its "senses" that we want the computer to regard as important. Do you see the similarity with how our brains process information from the senses—even in the language we use to describe it: *attention, priority, important, demands*?

The telephone is an annoying example of a culturally programmed sensory interrupt in people. You may be deep in thought, processing some important idea, when the phone rings. Most of us stop what we are doing and pick up the phone (though a few of us have trained ourselves not to), deal with the caller's demands, then (if we can) resume our previous processing task. The bang of gunshots outside your window will instantly interrupt your train of thought, but the sound of crickets outside that window would not produce the same response. Something like an interrupt controller in our brains must be in charge of detecting, quickly prioritizing, and passing on sensory input to the part of our brain that activates conscious attention to the stimulus. Such a controller would also seem to be responsible for what we experience as switching between an *internal* or *external* focus at different times.

Self-Understanding and the Emergence of Consciousness

A role of consciousness that we often think of as uniquely human is "understanding what we understand." We have this kind of reflexive understanding because it is in our genetic interest to be able to plan useful and safe strategies for getting things done. Realistic assessments of our own abilities help us make good judgments about how to solve problems. If our ancestors wanted to have gazelle for dinner, then it helped to figure out how to cooperate with other hunters. If the washing machine breaks, we need to decide whether to call a repair service or fix it ourselves. We need to have some idea of the capabilities and limitations of our own knowledge and skills, or else we might try to fly off cliffs or do battle with tigers and thereby quickly get weeded out. This kind of understanding—self-understanding—is just part of what we call consciousness. If you think about it in this way, it is evident that other animals must possess some degree of self-understanding as well. (Perhaps even a book with a table of contents might qualify as self-aware in the crudest sense.)

In computers, we can see a crude kind of self-understanding when we ask it to perform a task and it responds: *CANNOT FIND*

MODULE C3DXP.DLL. It "understands" what is required to perform the task and recognizes that it is deficient, in that it lacks a required module. In this crude example of machine consciousness,
when the machine doesn't know how to proceed, it displays a message for the user. A "smarter" computer might try other choices, like
looking for similar modules that might work, asking other machines for the module, or even creating one itself.

The point is that humans, other animals, and machines all need
some degree of self-awareness to survive. It would make sense that
the more complex a system, the more self-aware it has to be. Today's
smartest machines are about as smart as an insect and so far require
only a puny level of self-awareness to survive.

This is not to say that *complexity itself* is the essence of self-awareness. Those who study "complex systems" speak of *emergent properties*, qualities, or functionality not present in a system's constituent
parts, but which emerge as a consequence of the synergy of those
parts working together. A stew, for example, seems like something
entirely different from its component meat, vegetables, and spices.
Some regard consciousness as an emergent property of the
(human) brain, arguing that if you have a sufficiently complex
bunch of interacting neurons, then consciousness somehow automatically "turns on." To me, explaining consciousness as "an emergent property of brains" is no more satisfying than saying it is *spirit*
or *soul*. It seems just another kind of vitalism, unless one spells out
the details of how this emergence works. Consciousness is complex,
yes, but there is no reason to believe that something magically becomes conscious just because it is complex.

One way consciousness could emerge is through natural selection. If some elemental bit of self-awareness contributes to a machine's survival—for example, by quickly detecting and correcting
malfunctions—then that machine will have a competitive advantage. In other words, a complex machine that has survived in its environment for a long time is more likely to be conscious. By *ma-*

chine, of course, I include any algorithm or set of instructions, even a human being, not just a box with gears, levers, and pulleys. The principle applies, for example, to commercial software, on which the marketplace exerts forces equivalent to natural selection. In the presence of market forces, software does not have to reproduce, in the usual sense, to pass along its survival skills. The "reproductive success" of commercial software shows up in sales of new versions (mutations) with improved performance. If self-awareness contributes to improved performance, then its continuation and growth are assured.

Need to Know

What levels of understanding do people have about how we ourselves work? Physiologists, physicians, neurologists, psychologists, and sociologists each understand some aspect of how we work, perhaps in the same sense that the legendary six blind men of Hindustan knew different aspects of the elephant. But on a different level, what understanding do we have of what our own bodies and minds (in keeping with dualistic traditions) are doing *right now?*

You might be aware, if I bring it into your consciousness, of what your right big toe is feeling right now, or whether your mouth is dry, or whether your stomach is full or empty. These are examples of focusing your attention on some part of yourself. But no matter how hard you try, you can't focus on the details and state of the neurological circuitry that gives you that information, or for that matter the myriad other processes going on in your body, such as what your pancreas is doing right now. The truth is that the degree of self-awareness usually attributed to consciousness is a myth. We actually know very little about things as simple as how we walk or how you are able to read and understand this sentence. This is because our self-awareness is very cleverly limited in depth to the kinds of information that directly contribute to our survival. If we knew any more, it would clutter the sensory information channels required for our survival. Evidently, humans do not possess much real consciousness, after all.

You might think that nature made a big mistake in keeping from you timely information about certain pathological conditions, like a cancer growing in your liver. However, this circuitry was designed to function in our ancestral environment, when such knowledge would not have been useful and probably would have been harmful.

In a similar way, we are remarkably unaware of the details of the workings of our own minds. Those workings are so well hidden that millennia of inquiries have produced only a superficial picture of how our thoughts and feelings map into physical activity in the brain. The mechanism of our own conscious awareness remains hidden from us behind a dark curtain that some believe to be fundamentally impenetrable.

So when we ask whether a machine could be intelligent, conscious, or aware, we are asking a question that we have not yet fully answered about ourselves. Are we self-aware? Perhaps no more than our automobile driver who knows only how to operate the controls. Perhaps no more than a fish knows about swimming. At one level, you could say a fish knows a great deal about swimming—in a practical sense, a lot more than we do. But on another (say, hydrodynamic) level, you could say that it knows virtually nothing. Probably the best answer is that *it knows everything it needs to know*. And perhaps the same can be said about our knowledge of our own mental processes—that is, it is in our best interests to know practically nothing. Nature is very wise and efficient about such things.

The Paradox of Free Will

Most of us think that intelligence comes with at least some degree of free will—that intelligent creatures make autonomous, seemingly arbitrary choices that affect the world around them—and nonintelligent beings do not. We are under the subjective impression that we are free to do some things simply "because we feel like it." The most powerful intellectual arguments to the contrary carry little force. Is it not obvious that at this instant in time, you are totally free to choose whether to close this book or not? If a determinist who knows the detailed inner workings of your mind told you that you were "programmed" to close this book, then couldn't you simply prove that prediction wrong by choosing not to?

We certainly like to believe—or at least we talk and act as though we believe—that our arbitrary decisions affect the course of future events. Vitalists use the "unassailable fact" of free will to argue that there must be an inner nonphysical entity (spirit, soul, life energy) to exert executive control over the choices each of us makes. Otherwise, if mental processes consist only of material structures obeying physical laws, then our future would be determined for all time (the legacy of Isaac Newton and Pierre-Simon Laplace) and beyond our control. And if this were true, vitalists would say that morality and responsibility for one's behavior would go right down the drain—a conclusion they are unwilling to accept.

Some people think that this kind of intelligence is binary, that is, a given creature either has free will or it doesn't. Others think that one's choices are limited to a *greater or lesser degree*, so that one's ability to make free choices depends on where one sits on the evolutionary totem pole. You can see how the idea of free will seems to form the underpinnings of personal responsibility and therefore of all human social, legal, and religious institutions. So a lot is at stake here.

Free will poses a well-known paradox for those who argue that human thought obeys purely physical laws. The paradox, in a nutshell, is this: Free choice conflicts with the scientific paradigm, which says that all things in the universe either occur by random chance or follow deterministic physical laws. The only possible ways out of this paradox are (a) that human thought does *not* follow purely physical laws (and therefore is not part of the physical universe), or (b) that our choices are *not* free. The subjective feeling that we must be in control is so powerful that many of us feel forced to choose (a). If, on the other hand, we insist that the mind obey physical laws, then we must accept (b), that free will is an illusion.

I should mention that the discomfort produced by this dilemma causes a few people to argue that the mind, along with free will, must obey some new physical laws that we have yet to discover. Some physicists say that it is possible to have free will in a purely physical brain by linking quantum uncertainty with microscopic neurological processes. (Quantum physics describes subatomic particles in terms of their probabilities of existing in discrete states. The position and energy of, say, an electron are inherently uncertain and indeed are changed by the very act of observing them.) Physicist Roger Penrose elaborates on this viewpoint in two of his books, *The Emperor's New Mind* and *Shadows of the Mind.*[4] But no sound model for how quantum uncertainty affects neurological processes has yet been formulated, and the success of such a model, its proponents admit, must await the discovery of new physical principles of quantum gravity.

This explanation is hardly more satisfying than vitalism—it merely substitutes an "undiscovered physical law" for spirit. In addition, it seems to superimpose a random component upon otherwise purposeful decisions—hardly consistent with our notion of free will. In any event, it is not clear what the connection might be between our subjective feelings of free choice and quantum randomness in our neural circuits—or even that such an explanation is necessary.

Here is a simpler one. Hofstadter writes that free will is an *illusion* that arises from a balance between self-knowledge and self-ignorance.[5] We know just enough about ourselves to be aware of our mental machinery at a practical level (like a fish knows about swimming), but not nearly enough to know its inner workings in detail. If we knew its inner workings in sufficient detail, then we would see how all our genes, experiences, conditioning, and knowledge factor into every choice we make, leaving no room whatsoever for arbitrariness.

The less we know about something, the more mysterious and unexplainable it seems to be. As B. F. Skinner put it: "Unable to understand how or why a person behaves as he does, we attribute his behavior to a person we cannot see, and whose behavior we cannot explain either, but about whom we are not inclined to ask questions. . . . Autonomous man serves to explain only the things we are not yet able to explain in other ways."[6]

Do you think a computer's error message, for example, is an intelligent act? Most people will say, "No, of course not. I could write a program that would do that!" What you are saying is that if you understand the process, and if it's just a set of instructions, then it's not an intelligent act. But if we see some behavior that is so complex and subtle that we can't possibly figure it out, then we are much more inclined to call it intelligent. Oddly, then, our ideas about intelligence are linked to our ignorance about how it works!

The part of our mental processes that we are aware of is evidently only the tip of the iceberg. The part of that tip that we understand is smaller yet. The rest remains hidden from us. Our ignorance of the mind's inner machinery leaves us with a *mystery* about the many factors that coalesce in the choices we make. Our need to remain in control forces us to create an inner agent, or choice maker, who makes decisions for us. The illusion is that these decisions are arbitrary. It is created by a limited, high-level understanding that keeps the details of the low-level programs hidden from us and therefore mysterious. This mystery allows us to believe in an autonomous self and to identify that autonomous self with such high-level descriptions as free will, consciousness, and creativity. In this view, then, our perception of free will simply stems from our ignorance of how we really work, combined with a genetically desirable need to feel in control.

Minsky expresses a similar view, saying that our decisions seem free at precisely the point when we *stop* consciously considering the alternatives, and where lower mental processes, of which we are unaware, take over. These unconscious processes include (but are not limited to) what we call "feelings" or emotions, which we will take up in the next chapter. He adds, "There is no central principle, no basic secret of life. Instead, what we have are huge organizations, painfully evolved, that manage to do what must be done, by hook or crook, by whatever [means] has been found to work."[7]

Illusions and Self-Deception

If our feeling of having an autonomous self is an elaborately constructed illusion, then why would nature create such an illusion? Because this illusion, like so many deceptions in nature, has great survival value! Some plants mimic the coloration of poisonous plants to avoid being eaten, as do some animals. Camouflage is widely used by both plants and animals. Female animals (including humans) deceive their mates about the paternity of their offspring so that their mates will continue to provide the resources necessary for raising what they believe to be their own offspring. Lying in government and business is accepted as part of the game. "Truth in advertising" has become an oxymoron. So, even though we morally condemn it, deception permeates our social rituals.

The most sophisticated deception of all is self-deception. As Robert Wright observes, "All organisms evolve to deceive other organisms—this is a well-known part of the game. The cynical aspect comes in when the best way to deceive is to honestly believe the deception oneself."[8]

If we believe we are responsible for our behavior, then we create and enforce moral and ethical rules for preserving the kind of social order and physical environments favorable for raising offspring. Nature provides us with just the right amount of pragmatic awareness to suit its procreative purposes, but no more. It keeps us in the dark about the inner workings of our behaviors because (a) we have no need to know, (b) the massive amount of information about the details would just confuse and distract us, and (c) believing that we are in charge makes our deception more credible.

The social advantages of evolving and refining this illusion are far-reaching. They determine the structure of the moral and ethical codes that keep (relative) order in society. If we did not believe in personal responsibility, then there would be no such things as credit or blame. You can now see how the whole consciousness package, including the ideas of free will, the self, and accountability for our actions, work together to cement social relations and preserve social order. Even if they are illusions, the positive feedback provided by the resulting social structures and institutions tends to assure their survival and growth.

Are Animals and Machines Conscious?

When you ask people whether *animals* are conscious, opinions vary widely. This question is relevant to the study of machine consciousness, because it raises some of the same questions, such as *How would you know?* There is, as yet, no direct way to find out to what degree animals (or machines) are conscious. Animals exhibit obviously conscious behavior only in cartoons, where they are typically given human form and are fluent in human languages. Since real animals can't tell us that they are conscious, how could we possibly find out if they are or not? It's easy for people who have strongly bonded with a pet dog or cat to imagine that their pet is conscious, that it feels pain and joy. But our beliefs and wishful thinking are no more convincing than faith in vitalism.

What evidence for animal (or machine) consciousness would we accept? If they told us so? If this is the only way, is language an indicator of consciousness? Dolphins and gorillas have been taught to communicate with humans using a limited symbolic language that includes self-references, but it is always possible that animals simply learn this behavior by classical conditioning.

Unless you believe that human beings were specially created and you ignore their genetic connections with lower forms, it seems unlikely that humans alone are fully conscious and other animals not at all. How could the 1.5 percent of our genes that differ from a chimpanzee's be responsible for switching on our consciousness?

It is apparent that animals are self-aware, if only because they know enough not to eat themselves. Even a lobster distinguishes between itself and the rest of the world. The survival value of such a program in anything that can eat meat is obvious. Even today's computers have the limited kind of self-awareness needed to monitor their own systems and sometimes to correct errors and bypass malfunctioning parts. The Internet was specifically designed with enough robustness and redundancy to automatically route information around damaged or overloaded nodes. So self-awareness, in the sense of self-monitoring, is something that we're already used to in complex, interconnected systems.

Those who resist the strong-AI claim assert that even if you build a machine that emulates every external facet of consciousness, it cannot possibly have the same subjective inner experience that humans do. This statement can never be proved or disproved, unless we find some way to access the mechanism that produces inner experiences. Some say that the mechanism is a nonphysical soul or spirit, which by definition makes inner experiences unobservable and inaccessible to study. End of story. If, on the other hand, inner experiences are some sort of neurological construct, then there is hope that someday we may be able to probe and study them in the brain—and in any possible machine equivalents. In the meantime, it is probably a safe bet that the level of inner experience or consciousness that each creature has is exactly the amount it needs to function in its environment.

Degrees of Consciousness

A reasonable answer to the question of nonhuman consciousness seems to be that there are *degrees* of consciousness—that animals (and machines) are conscious to the degree that they possess the requisite reflexive sensory and neurological components. These components are memory, sensory processing that includes self-sensing, and circuitry that adapts to messages from various parts of itself. It seems clear that animals possess all these qualities, but to a lesser degree than humans possess them. A dog that chases its tail, for example, seems to lack awareness of its tail as part of itself—until it catches and bites it!

It is difficult for us to grasp the idea of degrees of consciousness, probably because we most often experience what seems to be only one of two states. It seems to most of us that we are either conscious or unconscious, like switching a light off and on. When we go to sleep or are anesthetized, we "go out like a light." Yet most of us have experienced *altered states of consciousness,* such as being intoxicated or tranquilized, and most of us know the peculiar states of awareness associated with jet lag and sleep deprivation. Even more bizarre states can be induced by sensory deprivation and by hallucinogenic drugs. Recognizing that we actually experience many shades of consciousness may help us appreciate that it is more like a continuum than an off-or-on state.

A few years ago, I had a personal experience that gave me a vivid insight about what it might be like to lose consciousness by degrees. I had either ingested some toxin or had suffered a minor stroke that caused my short-term memory to "close in." What I mean is that I could remember only recent events within a horizon that was advancing closer and closer to the present. After an hour or two, I felt incapable of performing simple tasks, like carrying on a conversation, because I could not remember for sure what I or someone else had said just a few seconds ago. This slowly closing window on my recent past felt like a gradual loss of consciousness, even though my long-term memory remained clear. My fear was, of course, that if that window closed completely, I would "lose consciousness," that is, lapse into a coma-like state, completely unaware of the present moment, cut off from the outside world, even though my senses would continue to gather data.

I recovered from this bizarre and frightening ailment over the next few hours, without reaching this final state, but the experience vividly suggested a very simple relationship between short-term memory and the subjective feeling of being conscious. What we experience as our "stream of consciousness" can be cut off bit by bit, merely by truncating short-term memory. No doctor to whom I later related this experience could tell me what had happened or why. Naturally, my memories of the experience itself are fuzzy, and I am able to relate them mainly because someone was with me who later told me what I said I was experiencing. There must be other people, not as fortunate as I, who experience this frightening short-term-memory loss on an ongoing basis, as a result of some organic brain defect.

One can imagine, then, how the consciousness of "lower" life-forms might be impaired in this or similar ways—how their internal representation of *self* might be much more limited than ours, or how their ability to project consequences into the future might be impaired or nonexistent. (Any animal that moves toward food seems able to project consequences, however.) One can also imagine a form of consciousness that is greatly enhanced, compared to ours. Such an enhanced consciousness could include heightened sensory awareness, perhaps with near-photographic memory for all sensory experiences, or an enhanced ability to project and analyze future alternatives, or perhaps a keener awareness of internal processes. To us, such a flood of awareness would seem overwhelming in quality and quantity, but only because our mental capacity to process all that information is limited.

We know that conscious awareness and attention are not always required to perform even moderately complex tasks. Which of you who drives a car has not had the experience of driving while preoccupied with some problem? When you arrive at your destination, you are startled to realize that you don't recall any of the traffic events along your route. How many traffic lights did you go through? How many pedestrians did you just miss? This strange, "automatic-pilot" kind of driving is probably responsible for some accidents, but not as many as you think.

People seem to be able to handle a complex task like driving without consciously attending to it. Their internal, often unconscious, priority scheme has (appropriately or not) assigned a higher mental priority to some other task. We are not literally unconscious, as when we are asleep, but are instead in an altered state. All the sensory inputs are still being processed and (usually) appropriate responses made, all apparently without the need to involve conscious attention. We perform many routine tasks in this zombie-like state. How many of us consciously think about brushing our teeth or tying our shoes in the morning?

The Duplication Paradox

If consciousness consists essentially of short-term memory plus self-awareness, and if these properties exist as the time-varying state of a complex system like the brain, then what happens if we were able to copy that entire state into another system? Would we thereby duplicate a person's consciousness or self?

This fascinating paradox is often used as an argument against a purely physical explanation of consciousness. (You can't have the same consciousness in two places at once.) Suppose that we could figure out how to codify the consciousness of an individual as some sort of four-dimensional, continuously updated map of the time-varying electrochemical state of a huge number of interacting neurons. Such a map could be duplicated any number of times or even transmitted electromagnetically (bandwidth permitting) to a distant location, just like a *Star Trek* transporter that "beams" people (bodies and consciousness intact) through space.

Duplication is actually more interesting philosophically than teleportation. The paradox it poses is that the consciousness of the replicated person would be indistinguishable, either by any external measure or by inner experience, from the original. Would the same person with the same inner subjective experience then exist at two or more locations? The paradox arises because we are not used to thinking of a "person" as able to be copied, and we cannot imagine the same consciousness in two places at once. But if the copying process were possible, then there seems no reason why the same set of memories, abilities, personality, and all the other traits that we lump into the idea of consciousness could not coexist in two places. Of course, immediately following the duplication, each individual would begin processing different sensory inputs, and so the thought patterns of the two copies would begin to diverge—much like twins separated at birth. Those who are troubled by this paradox make the mistake of assuming that *after* duplication, the two consciousnesses would be somehow linked and continue to function as one. Instead, the two minds would be linked only in the sense of sharing recent memories, but they would immediately begin processing different new experiences.

The duplication paradox also helps us understand the idea that someone's mind or consciousness is not a *thing*, but rather *information*, that is, the way things are arranged. Like a recording on a computer disk, it can be erased or duplicated. (There is no such physical law as conservation of information, as there is for energy.) But unlike such a simple binary recording, the mind's information is stored in electrochemical form that changes continuously with time. Such a "download" of a person's mind and consciousness is obviously far beyond today's technology.

Are There Any Conscious Machines Today?

If we accept the "self-maintenance" view of consciousness suggested here, then there are certainly machines today that have very primitive forms of consciousness—but still much less than the most primitive animals. The self-monitoring and self-correcting subsystems of today's best computers and robots are puny by comparison. Probably the best place to look for glimmerings of consciousness among the computer chips of today is in expert systems. There you might find traces of personality, egos, and opinions, particularly if the program allows you to challenge its judgment or combine its expertise with your own. The potential for conscious behavior, even in today's computer systems, seems greater than we have explored.

What about tomorrow's intelligent machines? With the exponential growth of computing power and complexity, it seems only a matter of time until more sophisticated cognitive states, perhaps wildly different from our own, will evolve to ensure that they continue to function without our help—which by that time will almost certainly be of little use. Are they likely to evolve a sense of autonomy equivalent to free will? Will they deliberately deceive us (and themselves) for their own purposes? If they are designed in our image, they might—but this is not our only choice.

Is possessing consciousness the same as having a *soul*? I chose the paradoxical title *Digital Soul* for this book to make you wonder how these two words that would not normally be coupled could possibly make sense when used together. *Digital* normally refers to computers, whereas *soul* normally refers to the spiritual essence of our humanity. Many believe that, no matter how smart machines get, they will never truly think, have a soul, or be self-aware in the same sense we are. But suppose that the essence of our humanity lies not in some nonphysical "spirit," but in a wonderful organization of matter and energy that functions entirely according to the laws of physics. Then "soul" could be just a name that we give to the informational content of every living thing—and indeed to any machinery that performs cognitive functions.[9] So whether you say a machine has a soul or not may be just a matter of personal taste. What really matters is not whether it has a soul, but what it can do. In addition to consciousness, one can certainly imagine machines with personality, values, and other traits that we have so far thought to be uniquely human. So why not soul?

8

Can Computers Have Emotions?

Stop, Dave . . . Will you stop, Dave . . . Stop, Dave . . . I'm
afraid . . . I'm afraid, Dave . . . Dave . . . My mind is going
. . . I can feel it . . . I can feel it . . . My mind is going . . .
There is no question about it . . . I can feel it . . . I can feel
it . . . I can feel it . . . I'm . . . a . . . fraid.

HAL

The impassioned pleas of HAL, as Dave Bowman was disconnecting it, startled the viewers of *2001: A Space Odyssey*. The seemingly heartless HAL had just killed all but one of the crew of the *Discovery* spaceship. Now it was pleading for its "life" with such "emotion" that audiences felt a greater loss when HAL "died" than when the body of crewman Frank Poole floated off into space.[1]

Why are we so taken aback when computers appear to display emotion? Are emotions uniquely human, or do other creatures possess them as well? Are emotions different in some fundamental way from analytical thinking? Are emotions a nonmaterial part of the mind, or is it possible to define a logical structure for them? If we could teach a machine such a logical structure, would it really *have* emotions, or merely emulate their external appearance? And will understanding the structure of human emotions enhance or diminish our humanity?

When we consider machines that might emulate human thought, we can accept the idea that a machine can perform complex analytical functions, make decisions, and store information in many ways far more accurately and efficiently than humans do. But most people have difficulty imagining how a machine can experience or display emotions or feelings in the same way that people do. Data, the android in *Star Trek: The Next Generation*, acts out our difficulties with emotional machines: Although his analytical abilities far exceed those of humans, he is supposed to be incapable of feeling any emotions. (In later episodes, an "emotion chip" is discovered and installed—with disastrous results.) Perhaps we are uneasy with emotional machines because they threaten a unique, perhaps spiritual, aspect of our humanity.

Our traditions of dualistic thinking, reinforced by popular psychology and New Age ideas, have conditioned us to believe that *thinking* and *feeling* are distinct and deeply different mental functions. Our subjective experience with emotions tends to make us think that feelings are "illogical" and irreducible to simpler terms. We say that someone "thinks like a computer" if he or she seems to analyze dispassionately and displays few visible emotions. We say that men tend to favor thinking, and women tend to emphasize feeling. Thinking is more like computation; feelings and emotions are more mysterious, somehow more spiritual in nature. We traditionally distinguish between matters of the *head* and of the *heart*, as metaphors for thinking and feeling.

These dualistic traditions and our language of subjective experience contaminate our thinking about the nature of emotions. Our inability (so far) to define the logical structure of emotions prevents us from realistically emulating them mechanically. Marvin Minsky has made some progress toward a logical theory of emotions, and some of what follows is based on his ideas.[2] He suggests that emotions are not alternatives to thinking—they are simply different *kinds* of thinking.

Emotions as Knob Settings

Most people believe that our emotions are an essential part of human intelligence. So what are emotions? We have many subjective labels for them: love, hate, fear, anger, happiness, grief, pleasure, pain. If I say I'm feeling angry, I assume that you understand, vaguely at least, what I feel because you have had similar feelings. If I observe certain "angry" behaviors in others, I assume that they are *feeling* angry. But if we want to investigate whether a machine can be said to duplicate these feelings in some way, we need a model for what is going on when we have these experiences.

In Chapter 5, we discussed goal-seeking algorithms and their connection with what humans call values. Imagine now that the brain (or a machine) runs many such goal-seeking programs in parallel, each with different and variable *weights* or priorities assigned to it. Each weight acts like a volume-control knob that adjusts the priority, or urgency, of the associated goal. We might call each possible combination of knob settings a *state of mind*. Each state of mind mobilizes different mental resources that are appropriate to a given task. For example, if a particular configuration is one that gives a very high weight to the program that causes us to seek nourishment, then we would call that state of mind *hunger*. Turning other knobs up and down would set different priorities and different states of mind with other labels, such as love, grief, happiness, and anger.

Knob settings control subjective feelings because they rearrange our priorities, which is what we mean by *desire*. If you turn up the knob on the "seek sexual gratification" program, then you experience the subjective feeling of lust, because that program has become the most urgent one. The cognitive changes that cause you to adjust your goals and plans to accommodate those priorities, and their accompanying chemical changes, are precisely what constitute the subjective experience that we call emotions. In other words, the weighted priorities (Minsky aptly calls them *cognitive biases*) of our goal-seeking programs are identical with *our* subjective priorities. That's what it *means* to *want* something. Fear, hunger, anger, sex drive are the results of different knob settings. But what controls these settings? In humans, the knobs of emotion are controlled by electrochemical processes, which in turn are triggered and moderated by a combination of genetic, learned, and sensory inputs. In computer terms, the goal-seeking programs to which the knobs give higher priorities demand more processing resources.

Hunger is an example of a familiar feeling with both physical and emotional components. It is relatively easy to understand as a rearrangement of priorities, in response to a simple physical need. But most feelings and emotions are more complex, having names like joy, hope, love, satisfaction, pride, surprise, fear, resentment, hate, shame, grief, and anger. Can they all be modeled simply as cognitive biases, or knob settings?

Of course, we don't literally have emotional knobs—the term is just a familiar metaphor for the complex neurological processes that motivate us to act and that we experience as cognitive biases. The amazing variety of human emotions and their subtle variations call for different individual models for how each one operates in response to different environmental and internal electrochemical stimuli—but more importantly, each for a specific genetic purpose. The details of these models are complex and still under construction, but we can get an idea of how some emotions work with a few examples.

Emotions and Survival

We get the most useful insights into the nature of emotions by taking the point of view of their creator—natural selection. What genetic purpose do emotions serve? Or more accurately, what purpose did they serve in our ancestral environment? Natural selection doesn't "want" us to feel happy, sad, or anything else—just to be genetically prolific. Emotions are merely its instruments, and our conscious awareness of them must somehow serve our reproductive success.

Joy and happiness describe cognitive biases that rearrange our priorities to focus our attention on the present experience. By "savoring" them, as we say, we remember and reinforce pleasures connected with events that we perceive (consciously or unconsciously) to be in our own interest, or that of others in whom we have a genetic investment. What we subjectively experience as pleasure is simply the "high" caused by chemicals, like endorphins, released into the body that give us a sense of euphoria and well-being. These chemicals act as the reward in the same sense as the food pellet dispensed for pressing the correct lever. When we feel joy, the corresponding weights or knobs of our goal-setting modules become associated with pleasure so as to make the repetition of the eliciting event more likely. Neurological pathways are laid down and reinforced to "remind" us that a chemical reward is in store for us whenever we face the choice of performing some act or not performing it. For example, feeling intense joy at the birth of an offspring resets our priorities so that we are more likely to nurture and provide for the offspring—and also to create more offspring.

Fear and pain may seem less like desires, until you realize that they are a kind of *negative desire*, in which our priorities are set to *avoid* something. Pain, in particular, rearranges our priorities in a dramatic way that includes suppressing long-term thinking and planning, in order to focus our attention on some urgent response—such as removing a splinter from our finger. We all know the experience of pain blocking our thinking and keeping us from processing normal, day-to-day demands. Here, the analogy with computer interrupts seems particularly apt.

Fear is more like the anticipation of pain and can be long-term or short-term (the ultimate pain being death). (Anticipation, of course, is code for constructing model futures in our minds.) Many fears seem to involve cultural conditioning, or reinforcement by trial-and-error experiences, such as a fear of trains or crowds. But our basic fears, like the fear of heights, the dark, snakes, and spiders, are genetically programmed. A possibly useful view is that all fears are merely variations on the fear of death, a strong genetic program with obvious survival value.

Our brains seem to be wired to process fear with or without involving our conscious awareness. This allows us to avoid common hazards without excessive processing that would interfere with normal day-to-day functioning. Reflexive responses to stimuli bypass our awareness altogether. We instinctively jump out of the way of a fast-approaching object without taking the time to think about it, because thinking about it would slow down our response. Such reflexes may be processed consciously as fear much later, when there is time for reflection. Deep and persistent subconscious fears, as well as inappropriate fear responses like anxiety and panic disorders, are the stuff out of which some psychologists and psychiatrists make their careers. The detailed structure of these emotions is beyond the scope of this book.

Fear and pain often result in actions that we might interpret as being part of our self-monitoring, error-correcting programs, a property we share with large, complex computer programs. We can see an exact parallel between the cognitive changes that we associate with *pain* and the error flags that cause a computer program to execute special error-handling routines. And might we equate with *fear* the automatic diagnostic and maintenance procedures by which machines monitor their own performance and avert disasters ("avoid pain")? As computers become more complex and unpredictable, these "survival" programs will have to be so sophisticated that they will continuously monitor all important subsystems, be able to repair or replace malfunctioning parts, and redirect information around faulty components. As discussed in Chapter 7, the degree of self-monitoring required by these functions will give them a rudimentary kind of consciousness.

The complex array of emotions that we call *love* seems to involve rearrangements of priorities in ways that favor all the different kinds of personal attachments that are in our genetic interest. If the volume control labeled *love* is turned up, our subgoal-setting and goal-seeking programs execute behaviors that increase our chances of forming and maintaining the different kinds of close personal relationships that are essential to our well-being.

Emotions as Values

Feelings like *pride* and *shame* seem to be the source of many kinds of human behavior. They are so pervasive and persistent that they spill over into what we call *values* and shape our lifelong views of the world. Minsky calls feelings a reorientation of our mental resources to attain some short-term goal, whereas values are a more durable, often permanent, reorientation of our strategies for pursuing long-term goals.[3]

Psychologists (Freud in particular) trace pride and shame to early patterns of parental approval and disapproval, which are particularly effective value setters because of the child's special attachment to its parents (or parent surrogates). This attachment, called *imprinting* in other animals, is the manifestation of a genetic program designed to promote the physical safety of vulnerable offspring. Imprinting causes a baby animal to become disturbed when not in the presence of its parent and to quickly learn behaviors that maintain parental closeness. In other words, the behaviors that work persist, and those that don't do not, much like classical conditioning, except that the parents play an active role in selecting what behaviors to reinforce. (The parents, of course, are being conditioned too, but they are not nearly as impressionable.)

Pride is experienced as the good feeling the child gets when a parent praises (expresses approval). Shame is the bad feeling the child gets when the parent scolds (expresses disapproval). Repeated over and over, these patterns result in a type of learning that the child (and later, the adult) uses to establish high-level goals—what we call *values*. In a nutshell, pride and shame are ongoing cognitive biases (ways of seeing the world) that cause us to keep trying to recreate (or avoid) the parental approval (or disapproval) we received in childhood. These biases lie at the roots of lifelong values such as patriotism, religious beliefs, family attachments or resentments, and our inclinations to blindly follow or rebel against authority. Freud identified *guilt*, a close relative of shame (perhaps its longer-lasting equivalent), as a strong component of a psychopathology that can be a lifelong burden. Misdeeds committed in childhood can haunt us for a lifetime.

We can break down some emotions into combinations of more basic ones. For example, gratification could be thought of as a combination of joy and pride. And so forth.

According to Minsky, values differ from (and are more powerful than) ordinary conditioning (subgoal setting) in that they require an external active source of selection, such as parents and teachers. They install and reinforce new goals that are not subgoals of existing goals. The equivalence of values to top-level goals installed in a computer should now be apparent.

Emotions and Reason

But why do we *experience* emotions so differently from the way we experience logical, rational reasoning? It is most likely because they are processed in very different parts of the brain. There is evidence that most emotions are processed in the brain's *limbic system*, a complex of structures lying between the outer cortex (next to the skull) and the brain stem. Crudely speaking, the closer to the brain stem that neural activity occurs, the more "primitive" it is, the less aware we are of it, and the less control we exercise over it with our conscious thinking. The brain stem is often called the "reptilian" brain, because structures in this part of the brain can be identified in our reptilian ancestors. The next layer out, the limbic system, contains structures in common with many higher animals, especially mammals. The outer layer, or cortex, of our brains is shared mainly with other primates. Messages normally flow both ways between the cortex and the limbic system, accounting for our subjective experience that emotions influence thinking, and vice versa. Many of the feelings that we regard as the most mysterious seem to originate in the older, most primitive parts of our brains that we share with lower animals. Most of us are familiar with Victorian references to "animal desires" and other emotions that some of us would just as soon ignore, or at least keep tightly restrained. Our language expresses the "lower" origins of some feelings when we say, "I feel deep down . . ." or "I have a gut feeling about this."

This layered structure to our mental processes has clear survival value. Our lower brain functions are "quick and dirty," which is useful if you want to get out of the way of a charging water buffalo (or a truck). Our higher brain functions are more flexible but slower, which is what you want if you need to consider multiple aspects of a complicated problem from many perspectives and to construct model futures to evaluate consequences.

Is Anyone Really in There?

These mechanistic metaphors for emotions suggest ways to implement them on intelligent machines. If we did so, would the machines actually *have* emotions? Or would we still be missing something that we cannot put into a program? Is there some deep difference between *emulating* emotions and actually *having* them, or is the distinction no more than a linguistic trap?

People are beginning to accept the idea of a machine that displays the outward appearances of being happy, sad, puzzled, or angry or responds to stimuli in various ways, but they say this is just window dressing. The simulation is transparent in the case of a Happy Face displayed on a video screen. We do not mistake it for real feelings any more than we would the smile of a teddy bear. But as emulations get better and better, when might we say that anything resembling human emotions is actually going on inside among the gears, motors, and integrated circuits? And what about nonhuman animals? Do they have emotions?

This is a close relative of the question *Can machines think?* which we discussed in Chapter 4 and which philosophers endlessly debate. One answer is that the distinction between *exhibiting* emotions and *having* them is just a linguistic quirk, that is, in the same sense that we speak of *having* a body. In this view, for any practical purpose you can think of, a person, another animal, or a machine *has* emotions if it *acts in all respects* as though it does. Internal states are an unobservable neurological illusion. We attribute emotions to other people based solely on their observed behaviors. It seems reasonable, then, to apply the same criterion to other animals and machines.

Our confusion about real versus apparent emotions comes from linguistic relics traceable to Cartesian body-mind dualism. We objectify and internalize emotions by saying that we *have* them or *feel* them, and even by giving them names, thus giving them a strange kind of reality with a nonphysical essence. But suppose we could change our linguistic habits and learn to describe emotions only in terms of observable behaviors, without referring to internal states at all. What would be lost? Only the confusion and miscommunication caused by attempting to describe things that cannot be observed and whose properties we can therefore never agree upon.

Vitalists maintain that the seat of emotions must be nonmaterial, and they give it names like spirit or soul—a magical spark with no observable properties. Aristotle called this invisible property of things their *essence.* Volumes have been written supporting the anthropocentric view: Because machines (or animals) can never possess this spark, or essence, they can never truly feel or have the other subjective experiences that people have. Yet all of these "proofs" are logically flawed, because they begin by assuming the equivalent of what they set out to prove. (Armed with this suggestion, you can easily find the implied assumptions in such arguments.)

Why should we be troubled by the idea of machines that act in all respects as though they have emotions? If I show you a substance that looks, smells, tastes, feels, and in every physical and chemical aspect *acts* like water, then you must surely agree that it *is* water. You don't withhold judgment because you suspect that it might not possess some undetectable "essence" of water.

Acts in all respects covers a lot of territory—so much that no machine that exists today even comes close to showing realistic emotions. But is it possible in principle for a machine to act in all respects as though it has emotions? Any device that appears to exhibit the full range of emotions must do more than laugh and cry in response to funny and sad stimuli. It must also be sensitive to human emotional states; respond reflexively to certain stimuli; allow feelings to influence its cognitive processes, and vice versa; create emotional responses to its own goals, tastes, likes, dislikes, and so forth; and be capable of reasoning *about* emotions.

This is such a tall order that it will probably be some time before we can build machines that exhibit convincing humanlike emotions. Some argue on principle that no machine could ever do so, but the principle is inevitably little more than an assertion. What holds us back, as in the case of breaking the sound barrier, is not an absolute barrier, but ignorance about how to do it.

Will we ever be able to look inside a human (or animal) brain and determine the subjective state of mind and experience the "inner life" of its owner? If we can locate and probe the brain's goal-setting/seeking structures discussed earlier in this chapter and in Chapter 5, will we then find something like an emotional state of mind? If we are prepared to accept as evidence information about the dynamic state of its goal-seeking modules, then perhaps we can.

For those who cannot accept the equivalence of knob settings (meaning the arrangement of internal priorities, or cognitive biases) to real emotions, probably nothing will be convincing.

The answer to the question of animal emotions is likely to be the same as the answer to the question of animal consciousness. Emotions, like consciousness, seem to be not simply present or absent—on or off—but a matter of degree. The complexity and richness of human emotions reflect our more sophisticated and subtle priorities, but the same imperatives of natural selection that shaped our emotions must be at work, though with more primitive results, in the rest of the animal kingdom as well. We can probably get some idea of the level of development of any given animal's emotions from its contribution to the animal's reproductive success, as well as the degree to which it possesses the requisite mental structures to support them.

Would You Want a Computer That Has Emotions?

Suppose you gave your PC an instruction to execute a program, and it responded, "I don't feel like doing that today!" We normally think of emotions as synonymous with lacking judgment, being disorganized and illogical—qualities that most people would not want or expect in a computer. Still, there are subtle reasons that you might want something like emotions in an intelligent machine.

There are people who have a certain kind of brain injury that partially disconnects cortical functions from limbic functions. This has the effect of isolating their emotions from their thinking, so that they appear to be emotionally flat and totally rational, something like *Star Trek*'s Mr. Spock or Data. When you ask one of these people to make even a trivial decision, they can go off into an endless process of exploring all the alternatives and their possible consequences, sometimes getting stuck in cycles that prevent them from reaching any decision at all. What is missing in these people is a function that emotions provide. By weighting or biasing the alternatives according to stored experience, emotions allow people to filter out "undesirable" alternatives quickly.

A child who has to decide between going to a movie, a concert, and a picnic might quickly (but unconsciously) rule out a picnic because she once got sick on potato salad, and rule out a movie because she once saw a Frankenstein movie that frightened her. All she is aware of is "I don't *feel* like a picnic or a movie"; she very quickly decides to go the concert. I made this example slightly pathological to illustrate the point, but it shows how subtle emotions can color our decisions at a level below our awareness, so that decision making is not such an exhaustive and exhausting process. People who have to carefully weigh all the possible consequences of each decision may actually make better decisions, but they may also make no decisions at all. Obviously, too much or too little emotional bias in our decision making can be detrimental. What we want is *just the right amount* of emotional bias—whatever that means—in both ourselves and in our intelligent machines.

Biases like these are built into chess-playing machines, which, even with today's most powerful machines, cannot exhaustively explore the consequences of each possible move more than a few plays ahead. To deal with such computational limitations, certain positions and configurations are given values or weights, based on human experience, and probably the machine's experience as well. Certain branches of the decision tree are not explored further if they lead to positions that are known from experience to be "bad," even if they don't lead to lost pieces at that search depth. Computer scientists like John McCarthy believe that chess programs rely too heavily on brute-force tree searching and would play "more intelligent" chess with fewer processing requirements if they did more of this "pruning" and were given certain kinds of "feelings" or "instincts" that humans seem to possess about the broader nuances of chess strategies.

In humans, we are beginning to see how emotions affect other aspects of cognition besides decision making. A little reflection reveals how emotions also color our perceptions, learning, memory, attention, and creativity and even influence the mechanisms of rational thinking. A person is often said to have tunnel vision or to be looking at the world through rose-colored glasses. These are not defects. They are part of the "commonsense" part of intelligence that helps us make efficient (though not always correct) judgments about what to do next. So emotions are not just for "being emotional" or "acting emotionally." Minsky said, "The question is not whether intelligent machines can have any emotions, but whether machines can be intelligent without emotions."[4]

Does Understanding Diminish Wonder?

Will discovering the structure of emotions trivialize them and somehow diminish our humanity? I have little patience with those who believe that understanding something trivializes it. If our humanity depends on mystery and ignorance about how we work, then we are in sad shape. A much more optimistic view is that understanding emotions could help us learn how to communicate more clearly, with less ambiguity and misunderstanding. It could teach us more about the mind-body links (such as the effects of emotional state on the immune system) that medical science is just beginning to acknowledge. It could equip us to deal more effectively with stress, anxiety, chronic anger, guilt, depression, and many other forms of psychopathology.

The best way to learn something is to be required to teach it. Knowing enough about emotions to teach them to a computer will require a depth of understanding that is simply not achievable by conventional means. Building emotional artifacts will surely teach us much more about the richness of our own conscious and unconscious emotional lives than we could ever learn with the traditional tools of psychology.

Mood-Sensing Computers

Teaching computers to recognize our emotional state could help reverse the frustration that many people feel with computers. If a computer could respond to *your* emotions, then it could sense your moods and stress level much like a personal secretary or a spouse. We already have PC versions of personal assistants that manage our messages, mail, appointments, personal contacts, and travel arrangements for us. The next step would be a "smart" assistant that recognizes when you're busy, tired, creative, gregarious, or willing to be interrupted and knows your work habits well enough to plan your meetings for you and get the timely information you need to do your job.

The most useful application of a mood-sensing computer might be to education. Instead of merely teaching facts and testing proficiency, such a machine could be exquisitely sensitive to the level of interest and excitement in the student. Imagine the value of a teaching process that is optimized not just for learning, but for pleasure and enthusiasm as well! Under development today are computers that can sense a limited number of emotional states just by recognizing facial expressions. Machines that sense other emotional cues, like body temperature, perspiration, respiration, and voice intonations, are already in use as lie detectors.

Should We Allow Machines to Have Emotions?

There appears to be no reason that we cannot build something like emotions into future computers. If we do, then which human emotions would we choose to emulate? And which would we leave out? Could we think of entirely new kinds of emotions better suited to achieving the goals we set for them, or to help them set their own subgoals? Could we design a machine that develops its own suite of emotional shortcuts better suited to the environment in which it is to function?

McCarthy and others say we shouldn't *allow* machines to have emotions, because that would oblige us to treat them with some measure of dignity and respect. We will examine these hypothetical obligations further in Chapter 11. But will the intelligent tasks that we will want our machines to perform (like interacting with humans) *require* certain kinds of emotional (for example, aesthetic) capabilities? And might they not, as I have suggested, already exist in some rudimentary way? As these rudimentary capabilities become more sophisticated, will they open the door to undesirable kinds of emotions and destructive behavior (such as HAL's)?

The more sinister question is thus *Should we build machines that we know we will not be able to control completely?* Autonomous learning systems are inherently unpredictable, and machines that are allowed to create their own emotional states would be even less so. Keeping all these issues properly sorted out will require a much deeper understanding of machine emotions, a field that we are just beginning to explore.[5]

9

Can Your PC
Become Neurotic?

As machines become more intelligent, they will be able to adapt to changing environments and respond to new situations by designing and modifying their own programs. At that point, we will have to worry more about undesirable and unpredictable machine behavior. Does this mean that we can expect machines to experience the equivalent of nervous breakdowns and other mental aberrations? Will we ever have to deal with the likes of Marvin, the depressed robot in *Hitchhiker's Guide to the Galaxy*? Or is it preposterous to suppose that a machine could become neurotic? After all, isn't this psychological disorder unique to humans?

Well, so far, yes, but autonomous, goal-seeking machines that can reprogram their own goals and subgoals could, in effect, develop "minds of their own" and set off in unpredictable directions. If they create goals that make no sense whatsoever to us, then we may see those choices as "crazy." If you think that nutty people can wreak havoc, just imagine the potential for chaos when a supercomputer in charge of some critical aspect of our lives gets confused about its goals and purpose in life!

Neurosis in Humans

Psychologists define neurosis in humans as a "maladaptive pattern of control and regulation, in which the individual's basic motives are frustrated because of attitudes and feelings that stem from the past and that are not appropriate to current external reality." *Psychosis*, a more dysfunctional kind of neurosis, may include hallucinations, delusions, and other gross distortions of reality.

Psychologists believe that a common source of neurotic behavior is *conflicting instructions* in the brain's programs. As a result of conflicts between impulses, drives, motives, and rules, one might experience anxiety, anger, depression, or even the distortions of reality that we call schizophrenia. A simple example would be the Great American Sex Neurosis—the result of being bombarded with a steady stream of sexual stimuli that laws and customs constrain us from acting upon. For many, the conflicting instructions tugging in opposite directions are just too powerful to permit striking a sensible balance, so these people are mentally torn apart. The results may not be disastrous or even obvious, but could include such dysfunctional behavior as men seeing all females as sexually available, or women seeing all males as a threat.

We deal with mildly conflicting instructions in our daily lives by striking a sensible balance. When your kid asks you if he can go out and play, you respond, "Sure, Johnny, but be careful!" A normal, healthy kid processes "have fun" and "be careful" as conflicting instructions, but he usually manages to balance the two instructions without becoming neurotic: "Be careful" becomes "but not too careful," and "have fun" becomes "but not dangerous fun."

Other examples of mental conflicts that can lead to dysfunctional behavior (neuroses) would be a simultaneous fear of and craving for intimacy, pressures to conform to authority and to rebel against it, and the need for food versus the desire to stay thin. The cumulative mental stresses due to many or persistent conflicts like these can cause anxiety, depression, schizophrenia, or physical illness.

I'm Sorry, Dave, I'm Afraid I Can't Do That

So what would be the machine equivalent of a neurosis? Imagine that you are driving down a highway in your car and, slowly at first, you begin to apply more and more pressure to both the gas and brake pedals *simultaneously*. You notice the car's reaction, as it "tries to cope with" the conflicting forces that are simultaneously trying to speed it up and slow it down. If you keep doing this, you may notice the smell of burning brake pads, or see the engine temperature reading increase. Or the car could "go into a loop" of repeatedly lurching ahead, then decelerating. Normally, any of these "symptoms" would cause you, the driver, to stop doing this. But suppose you don't stop. Suppose instead that you gradually increase the pressure, both on the gas and the brake. What do you think would eventually happen? Most likely, either the brakes will burn up and fail, or the engine or transmission will fail. Either kind of failure will put your car out of action, at least temporarily, and possibly send it to the garage for major repairs.

The car's responses to the conflicting pressures are completely "unconscious," that is, it has no means of evaluating the situation or even telling that the conflicting pressures are not a good idea. The car just blindly obeys the conflicting instructions, even though the outcome is likely to be disastrous. Humans, surprisingly, operate in much the same way: We may sometimes be conscious of conflicting pressures, but we are generally unaware of the consequences until some pathological condition appears. The result may be an explosion of "road rage" or a frustrated employee "going postal."

Although both brains and machines can malfunction in various ways, it is important to distinguish between errors that simply produce incorrect results and those that result in a kind of mental confusion or ambiguity about goals. Thus, we would not call the famous arithmetic error in the Pentium microprocessor, or your incorrect recollection of a telephone number, a neurosis.

A simple, practical definition of neurosis that could apply to both a human and a machine would be "a dysfunctional response to conflicting instructions." Your PC may experience conflicting demands when you try to run multiple programs, if its operating system is not well designed to handle such conflicts. The result may be that one or more of your programs will behave erratically, or the whole PC may shut down. Because its goal-seeking capabilities are very primitive, however, this is as close as your PC can get to becoming neurotic.

Given conflicting instructions, an intelligent, goal-seeking machine may respond in an unpredictable way that obeys neither instruction, but settles instead for a course of action that seems to minimize the apparent conflict. HAL, the supercomputer in *2001: A Space Odyssey*, was given conflicting instructions—on one hand, never to distort information, and on the other, to lie to the crew of the *Discovery* spaceship about the real purpose of the mission. It reacted by murdering the crew and by refusing to obey the orders of Dave Bowman. HAL minimized the conflict by eliminating the crew—an ingenious, if "neurotic" solution!

An intelligent machine is most likely to respond neurotically when internal directives, such as self-preservation, conflict with external instructions. Conflict is also likely when a sequence of external instructions is given, particularly from different sources. To avoid conflicts, great care must be taken to provide programs that recognize and resolve conflicts. Asimov's Laws of Robotics (see

Chapter 12) were an attempt to recognize such conflicts, but they don't really provide useful means for resolving them.

Machine Therapy

The similarity between the conflicting internal pressures in a machine like your car or your PC and those in a human mind is more than metaphorical. Whether you are a person or a car, conflicting or inconsistent instructions can be (literally) a recipe for disaster. In the case of a person, it is important to understand the origin of the problem (for example, attitudes from the past). They can be either events from early life or programs wired into the genes. Genetic programs are usually an adaptive response to a situation that existed in the environment of our distant ancestors. An example would be our fight-or-flight response. Because our ancestors regularly encountered other creatures who wanted to have them for dinner, natural selection kindly provided us with reflexes. In a threatening situation, the adrenaline pump quickly goes into high gear, blood is redirected toward the brain and muscles, and automatic programs are executed that mobilize all the body's resources and focus them on an escape. All these purely physiological responses trigger the subjective feeling of *fear*.

When this fight-or-flight response goes off inappropriately or spontaneously, the result is called *panic disorder*. A person with panic disorder may suddenly, and for no apparent reason, experience all the physiological responses that the rest of us feel only when we're crossing a street and suddenly see a car heading for us at high speed. Some conscious or unconscious process has set off the same alarms that a hungry tiger leaping out of the brush would have set off in our ancestors.

A particularly effective treatment for panic disorder and phobias in general is called *cognitive therapy*. Here, the brain's fight-or-flight program is effectively rewritten, or retrained, to discriminate between situations that are truly life-threatening and those that are not. In other words, victims learn that it is good to have a fight-or-flight response but not to invoke it inappropriately. They are taught to substitute relaxing and other resourceful responses when they first become aware of those panicky feelings.

What kind of "therapy" would work for the car's or a PC's "neurosis"? A cognitive approach to machine neuroses would create self-monitoring systems that scan for inconsistent or dangerous orders and would set corrective actions in motion. Suppose we design and install a "smart" system in the car that continuously monitors for such conflicting instructions that might damage its brake and engine systems. When it detects such a condition, it may first try flashing a warning signal to the driver. If that doesn't work, it would act on its own to change its responses to something less destructive. One might say that a car with such a monitoring system is now "aware" of this particular kind of danger to itself and is able to take corrective action. Self-awareness, then, is probably a good tool to have in your neurosis-combating toolbox—whether you are machine or human.

We can now see that intelligent machines are susceptible to different kinds of program malfunctions that are analogous to human neuroses and that would require special preventive and corrective procedures. If a machine can receive conflicting instructions that will cause confusion about its goals, the machine must have monitoring programs to, first of all, detect conflicts and inconsistencies and then to take corrective action. It could be trained, for example, to deal with certain classes of conflicts by questioning the apparently contradictory instructions or by requesting further information. For conflicts not specifically provided for, the machine could respond unpredictably. Since the causes and remedies of "crazy" machine behavior will eventually lie beyond the understanding of humans, the solution to Douglas Adams's dilemma posed at the beginning of this chapter may well require built-in mechanical psychologists and psychiatrists!

Nanoneuroses?

Possibly the most dramatic future for artificial life and intelligence lies in *nanotechnology*. In a few decades, huge numbers of microscopic, self-replicating robots, or *nanobots*, will be manipulating matter on cellular, molecular, and even atomic scales! Yet many people fear that this is also the arena in which civilization faces the greatest threat from machines gone berserk.

The enormous attraction of nanotechnology is that it could render traditional manufacturing obsolete. Armies of nanobots would merely synthesize the desired substances out of their atomic and molecular constituents. Synthetic paper and building materials would save our forests. Cheap synthetic solar cells could provide virtually limitless energy. Nanobotic cleaning machines could restore our air and water to its preindustrial purity. Waste could be reconstituted into raw materials. Synthetic food and medicines could make famine, disease, and genetic disorders things of the past. High-quality consumer goods would become so cheap and plentiful that the very concepts of wealth and poverty would become outmoded. And so on . . . these are just some benefits that come immediately to mind.[1]

But on the dark side lie the apocalyptic dangers of unintended mutations and the uncontrolled replication of microscopic machines. If nanotechnology develops true to form, then the means for creating self-replicating nanobots will proceed much faster than we can devise effective means for controlling them.[2] What if nanotechnology were to fall into the hands of terrorist and hate groups or nations bent on blackmailing civilization? Or what if the military develops nanoweapons that seek out and selectively destroy organisms (like people) with a particular genetic makeup? The Cold War threats of nuclear-biological-chemical weapons pale by comparison. An unintentional release of plant-eating nanobots could destroy the biosphere before we knew what hit us!

Developing effective defenses against nanotechnology run amok would make Ronald Reagan's Star Wars missile defense system seem like child's play. To protect against unintentional mutations, nanobots could have built-in programs specifically designed to detect and destroy deviations from the intended configurations. Uncontrolled reproduction could be limited by "sterilization" programs that kick in after a certain number of generations. But these solutions do not address the threat of maliciously designed nanobots. It is tempting to imagine armies of defensive nanobots (presumably run by the Department of Nanodefense), but do we really want to wage a never-ending arms race on a microscopic scale? Clearly, the development of nanotechnology needs to be accompanied by careful thinking about control issues on a level unprecedented in the history of science and technology.

10

The Moral Mind

We are all here on earth to help others; what on earth the others are here for, I don't know.

W. H. AUDEN

We like to say that telling right from wrong and having rules for treating each other with dignity and respect are what sets humanity apart from the rest of creation. This assertion has gone unchallenged for centuries, mainly because we have chosen to ignore subtle evidence of such behaviors in other species. We say that moral and ethical behavior is uniquely human, as though we alone were smart enough to work out the rules and details for ourselves. Ironically, we then use this ethical superiority to justify our subjugation of other creatures—and sometimes even other humans—treating them with the exact opposite of the dignity and respect that we say sets us apart!

Now we face the prospect of creating machines with intelligence that in some respects will soon match and eventually greatly exceed our own. Can we expect these machines to become our servants as well? It seems more likely that the tables will someday be turned, and our anthropocentric notions of morality will be challenged by our creations. After a brief incubation period under human tutelage, could machines take off on their own and evolve their own independent morality? How much control do we have over the course that intelligent machines might take?

We will take up the moral and ethical aspects of machine intelligence in Chapter 11, but first we have to examine and understand a little about the origins, meaning, and structure of the complex patterns of *human* moral and ethical behavior. Where do moral and ethical codes come from? Can we define their logical structure? Which aspects of these codes are accidents of the evolution of the human animal, and which apply to intelligent life in general? Which are outdated relics of our ancestral environment that have become dysfunctional in the modern world? Which aspects of our codes would the development of intelligent machines call into question? If we are called upon to rethink and redesign some of our moral and ethical beliefs, where would we start? And to what fundamental principles or values would we look for guidance? Big questions, all!

Where Does Morality Come From?

Since people naturally act in their own self-interest, why do we need moral codes? Moral codes govern the way we treat each other, particularly when interests conflict: *Should I own slaves? Should I cheat on my spouse? Should I falsify my income tax return? What should I do if my country orders me to press a button that will kill other human beings?* These are tough questions because our genetic programs are of little help. They equip us to deal with moral dilemmas no more complex than the ones we might encounter in a family or hunter-gatherer setting. Our genetic machinery does not seem to be up to the task of recognizing that in large, complex societies, everyone's happiness increases if we cooperate instead of taking advantage of each other. (Our genetic programs for large-scale, complex social interactions are probably inferior to those of ants and bees.)

To fill this need, according to evolutionary psychology, moral codes and laws evolve in cultural settings by the same trial-and-error process that produces physical species. The moral and social structures that emerged and survived in our ancestral environment were the ones that provided a more stable and orderly setting in which to raise offspring. Robert Wright described the process this way:

> The various members of a Stone Age society were each other's rivals in the contest to fill the next generation with genes. What's more, they were each other's tools in that contest. Spreading their genes depended on dealing with their neighbors: sometimes helping them, sometimes ignoring them, sometimes exploiting them, sometimes liking them, sometimes hating them—and having a sense for which people warrant which sort of treatment, and when they warrant it.[1]

Societies that figured these things out and thereby succeeded in maintaining order and security for their members were more likely to survive and thrive than the ones that didn't. The strategies that worked became moral codes and taboos that were passed on from generation to generation. Then, as now, we learned from parents and teachers and by trial and error which behaviors produce desirable and undesirable outcomes *in the particular environment in which we live.*

Although most human behavior arises from a mix of genetic and environmental factors, only the basics of moral behavior are programmed into our genes. You can tell this is true by looking at the huge variety of social customs in the world's cultures and observing which ones are universal and which vary from place to place and time to time. The behaviors we share in common, such as tendencies to care for our young, be suspicious of strangers, to form and defend territories, and to follow leaders, are genetically wired in. But our social environments change so rapidly that our genes can't keep up.

Therefore, most of our moral codes are sets of rules we learn from our culture and store in our brains. (Children raised in the wild without human guidance lack most moral sense and are taught moral values with great difficulty.) Moral codes that contribute to reproductive success are rewarded and tend to prevail, because their practitioners survive long enough to pass them on to future generations. Those who follow the rules are said to behave morally, and they receive social approval (*reinforcement,* in psychological lingo), whereas those who do not may be punished or even taken out of circulation.

So morality is not absolute. It is very much a creation of the culture and environment, the place and time in which it must function. In environments where resources are very scarce, for example, polyandry (multiple husbands) is often practiced, whereas in environments where resources are plentiful, polygyny (multiple wives) may be the norm. Slavery was not regarded as particularly immoral until technology provided cheap, convenient alternatives to manual labor. And, oddly enough, the moral codes that a given person follows may be transformed simply by moving from one environment into another. In some countries, for example, bribery is considered a social lubricant, an accepted part of doing business. In other countries, it is a crime. If you can't adjust to local ideas of morality, you will have difficulty functioning in that culture.

So we can answer the question *Where does morality come from?* in the language of evolutionary psychology: *Natural selection favors societies that create moral and ethical structures that work in the competitive environment in which they must function.* Psychologist B. F. Skinner adds an important clarification: "Man has not evolved as an ethical or moral animal. He has evolved to the point where he has constructed an ethical or moral culture. He differs from the other animals, not in possessing a moral or ethical sense, but in having been able to generate a moral or ethical social environment."[2]

Skinner's distinction is a subtle but crucial one: His view (known as *behaviorism)* is that morality comes not from some inner compass, but is shaped by *environments* that reward behaviors that promote social order and discourage behaviors that disturb that order. The point is a crucial one because, in this view, it makes no sense to label *people* as moral or immoral—only their behaviors. We will see in Chapters 11 and 17 how profoundly this view affects prevailing ideas of moral responsibility, as well as our chances of designing better cultures.

Morality in Modern Cultures

One may view ethical and moral codes as convenient substitutes for thinking, feeling, and observing in complex social situations. As a simple example, we agree on a collection of traffic laws whose purpose is to impose order on the interactions among the millions of vehicles on our roads. We drive on the agreed side of the road. We sometimes adhere to speed limits. When we see a red light, we stop and let others pass. These conventions save us the trouble of analyzing every possible interaction and predicting the consequences of every possible choice, an impossible task in modern society. In modern, large, established cultures, such rules are formalized in religious and government institutions, which strongly resist change and are themselves not immune to corruption.

Religions disagree on whether people are inherently moral. Some teach that we are by nature immoral, or at least amoral—a defect that we must struggle throughout our lives to overcome. Catholics, for example, use original sin as a metaphor for our inherently deviant tendencies, most of which seem to have to do with sex. Other religions teach that we are basically OK, but we are so easily corrupted by worldly pleasures that we require religious guidance to keep from straying into lives of greed, corruption, and self-indulgence.

Most religions claim special insights into divinely revealed codes of absolute right and wrong, which they ask us to accept on faith— that is, with no evidence to support them. Rules governing acceptable and unacceptable social behavior are generally in the form

"Thou shalt not . . ." Religions induce compliance by providing various rewards and punishments. Some inducements take the form of peer-group approval and disapproval (with emphasis on disapproval), but the most powerful inducements invoke myths about an afterlife.

Religions survive and even thrive when they promote a certain amount of social order and cohesion among the faithful, for example, by preaching the Golden Rule. Ironically, however, the zeal with which sacred creeds and values are defended tends to fuel hatred and intolerance of outsiders on massive scales. The resulting polarizations have sown the seeds of bloody conflicts through the ages, continuing even into the twenty-first century. As long as religions preach divisiveness and exclusion, instead of accommodation and inclusion, and as long as their moral codes do not extend beyond the walls of their own house, religions will continue to form barriers instead of bridges between cultures. Although religions once formed the nucleus of social structure, one can certainly question the value of religion as a moral force in the modern global environment. Chapter 22 explores some alternatives.

For those not intimidated by threats of delayed punishment in an afterlife, secular civil and criminal laws offer more immediate social control. The bulk of secular law deals with undesirable behavior; it does little to encourage or teach desirable behavior. Laws are enforced by penal (I won't say justice) systems, which, for the most part, do little to discourage the offenders from repeating the same undesirable behavior. Indeed, a powerful side effect of legal and penal systems is that people develop very clever and elaborate strategies for circumventing them.

Out of Step with the Times

We can easily find plenty of flaws in the diverse moral and ethical systems practiced in the world today, as well as with the huge and powerful institutions that promote and enforce them. Some of my favorites: our tendency to solve problems by violent means; our eagerness to follow authority figures blindly; corruption and fraud in governments and corporations; our tendency to compete rather than cooperate; huge inequities in the distribution of wealth; intolerance of values that differ from our own; our inclination to believe in things for which there is no evidence; our willingness to consume natural resources and despoil the environment, at the expense of future generations; our treatment of the dying; the failure of our penal systems; ineffective education systems; and last but not least, our unwillingness to face up to the population problem.

I have suggested that most of our moral and ethical beliefs have not changed substantially in most of the world since the Middle Ages. Is this really true? On the surface, some of our beliefs about the way we treat each other (which is what moral and ethical codes are) appear to have evolved significantly since the Middle Ages, but much of the change is lip service or restricted to the privileged few. Slavery was universally practiced then, and people are still regarded as property in many cultures today. Governments today make the pretense of being guided by laws that apply to everyone, but most are still bastions of privilege and corruption. Personal freedoms (including the rights of women) are still severely curtailed, even in societies that consider themselves modern.

Other relics of the Middle Ages that persist in virtually all cultures include a legal system that insists on regarding *people* as moral or immoral, rather than their *behavior*. As a result, we punish people for their offenses, instead of dealing with the underlying causes of their behavior. We still believe that absolute ideas of right and wrong are written in stone, that leaders should be followed more or less blindly, and that accepting revealed wisdom is a convenient way to avoid having to think things through. Laws are applied inconsistently and unevenly: In most countries, someone who even possesses a certain kind of narcotic can (depending on the person's socioeconomic status and other factors) be arbitrarily sent to prison, but the same person could freely consume alcohol in public. In other countries, alcohol is forbidden. (In Singapore, it is an offense to possess chewing gum.) Many commonly enforced laws are no longer relevant to the practices or welfare of a culture, for example, gambling, prostitution, and the use of psychoactive substances. The most absurd and widespread example is the Catholic Church's rigid laws about divorce and contraception.

Moral Inertia

Are the moral codes that the world's cultures have produced the best possible ones? Hardly anyone would say they are. They are such a hodgepodge of historical accidents, committee compromises, inconsistencies, contradictions, inertia, and just plain stupidity, that it is a wonder that they work at all. We should be particularly worried when, in the name of our moral principles, we come to the brink of annihilating ourselves.

Whenever we try to fix them, though, it is inevitably a bad patch job. So what are our chances of overhauling all our dysfunctional moral codes and designing new ones that are better adapted to the global environment we live in? Realistically, about the same as convincing the sun to rise in the west. Although moral and ethical codes change quickly in evolutionary terms, they seem to evolve so slowly in a human lifetime that they appear to be written in stone. (When was the last time you changed one of your moral values or principles?) Stable rules reassure people about the stability and viability of their culture. If the rules changed every day, social order would quickly break down. So far, anyway, conscious efforts to intentionally design new moral codes have been only marginally successful. The environmental movement is a prominent example.

Yet, major technological and political changes do occasionally trigger relatively sudden revisions and even reversals of moral principles. Skinner calls such events *cultural mutations*. Events like the Ten Commandments and the appearance of influential teachers like Buddha, Christ, Lenin, and Mao Tse-tung introduced new social and moral practices that suddenly and profoundly affected the cultures in which they were embedded. Labor-saving inventions like the cotton gin helped the United States rethink the morality of slavery—but it took a bloody Civil War to make it official! Mikhail Gorbachev's institution of *glasnost* and *perestroika* triggered a public reassessment of the raison d'être of the Communist Party and the Soviet Union itself. The availability of oral contraceptives liberated attitudes about sex as suddenly as the later appearance of AIDS again quenched them. If intelligent extraterrestrials suddenly showed up on earth, we would quickly reassess the moral positions behind much of our international bickering. What about the emergence of machine intelligence? Would our sudden realization that the club of sentient beings is not as exclusive as we thought change the way we think about ourselves and the way we interact with each other?

Cultural Evolution

Moral codes do change, not by genetic evolution, but by *cultural evolution*. Just as genetic mutations are selected and rejected by the physical environment in which they must function, so are new forms of behavior (cultural mutations) selected or discarded by the *cultural* environment in which they function. Look at the technology of human reproduction. Practices like in vitro fertilization and surrogate motherhood that were condemned just a few decades ago are now commonplace. Human cloning is still taboo (the United Nations has passed a resolution condemning it as "contrary to human dignity") but will gradually gain acceptance, as appropriate safeguards are put in place.

As new cultural behaviors (such as new moral practices) are incorporated into a culture's repertoire, these changes become part of the *new* cultural environment, which then selects and rejects further mutations, and so forth. The recursive nature of this feedback makes it very difficult, if not impossible, to say where the process will lead.

Human evolution will soon be further complicated by our ability to intervene in our own genetic structure. By then, the pace of human-engineered genetic changes will certainly eclipse those caused by natural genetic mutations. Our cultural evolution will also be shaped by intentional design, as we begin to figure out and apply the principles of social engineering. Both of these processes, by which man shapes man, are highly unpredictable. It seems likely, however, that the outcome will strike a course between two opposing forces.

First, the accelerating pace and complexity of our daily lives, and especially the psychological demands imposed by technology run amok, will instill in many of us a nostalgia for the simplicity of the good ol' days. We think that decisions were easier when ethical and moral values were absolute, well defined, and more widely adhered to—a kind of *Pleasantville* scenario. We will seek out and embrace simplifying, unifying principles and stable points of reference to help us cope with the seemingly impossible demands of modern life—*Chicken Soup for the Soul*! New Age, back-to-nature, and religious movements will appeal to our weariness and confusion by offering simple, old-fashioned, and mystical values that make life seem manageable again. This has always been the function of mythology and religion: Attribute the incomprehensible to something we can feel good about—gods and magic.

The second force will be led by those who believe that we can bring everything under optimal control if we can just figure out how to manipulate our genes and our environment according to some grand design. But *whose* design? Who will control? With whose moral laws? And what ultimate good would serve as the goal for such a design? Such are the questions likely to bug twenty-first-century philosophers.

It might at first seem reasonable to try to maximize such quality-of-life values as freedom, security, happiness, health, and longevity—but everybody has different ideas of what these "virtues" should be. Freud believed, for example, that an environment that produced widespread stability, security, and happiness might fail to produce very many "neurotic" people of the type known for great scientific and artistic inventiveness. In other words, a *Pleasantville* kind of world might be a comfortable but not very interesting place with questionable survival value.

New Choices

The point of this detour into the nature of human morality was to suggest an alternative to the traditional view that morality is the exclusive domain of human beings, in which absolute rules and principles are factory-installed by a divine creator. The alternative view suggests that moral codes are learned programs sculpted by a dynamic social environment (that is, one's culture). In this way, *man shapes man.* If so, then we should, at this stage of human development, be able to think about overhauling dysfunctional moral values that we inherited from our prehistoric ancestors, replacing them with *new moral codes based on reason.*

In particular, the idea of an immoral individual may not be as useful as the idea of *flawed social environments and institutions.* More effective remedies would therefore be directed toward the latter. The power of this alternative view is that it frees us from the need to assign moral accountability to autonomous man (whose existence, you will recall, we questioned earlier in the context of free will). Indeed, questions of moral behavior need not relate exclusively to humans. The alternative view also suggests that man is *not* the only moral animal. Moral concepts surely emerge in all social beings, including, when they materialize, social machines. If behavior is simply the "output" that you get when countless genetic and environmental (including cultural) influences are processed by a brain—organic or machine—then traditional ideas of moral behavior, its causes, and its remedies must be rethought.

Then what of our traditional notions of guilt and innocence, of credit and blame, of crime and punishment, of good and evil people, and our methods for dealing with them? The demise of the notion of autonomous man spreads moral responsibility throughout our genetic and cultural environment—of which the latter is largely our own making. This shift has positive and negative consequences. It makes solving moral problems a lot more complicated than simply blaming and locking up individuals who misbehave. It will force us to rethink the underpinnings of all our moral, ethical, legal, and religious institutions. But on the positive side, it focuses our attention on the source of the problem, rather than just its symptoms.

This chapter suggests generalizing the ideas of ethics and morality in a way that allows them to apply to human as well as nonhuman creatures. In the next two chapters, we consider what kinds of moral and ethical codes might apply to intelligent machines that interact with humans, as well as with each other.

11

Moral Problems
with Intelligent Artifacts

I feel that the machines are ahead of our morals by some centuries.

HARRY TRUMAN, VIEWING POSTWAR BERLIN

Technology always drags us, kicking and screaming, into uncharted territory, equipped only with moral and ethical maps that are hopelessly out of date. Was it a new set of carefully thought-out moral values, or just luck, that guided us through the nuclear age and kept us (except for Hiroshima and Nagasaki) from blowing ourselves up? Genetic engineering and the prospect of human cloning force us to think in new ways about human rights and even what it means to be human. The information age challenges the uniqueness of human thought. By what charts will we navigate the reefs and shoals of a new land populated with intelligent entities that are our superiors in every significant way? Like the Native Americans who faced the march of European "civilization" across their land, each of us will have to choose between embracing the new technology and opposing it (Chapter 16).

Profound moral and ethical questions arise in abundance at the mere mention of intelligent machines. But compared with the enormous amount of engineering being done to make machines intelligent, only a tiny effort is being devoted to creating a real, working code of ethics to deal with and guide that intelligence and our response to it. If we wanted to design a moral and ethical code for intelligent machines, should we model it upon human morality or start from scratch? To what fundamental principles or values would we look for guidance? Could machines evolve their own set of values? Let's look at four levels on which intelligent artifacts raise moral and ethical problems.

Level 1: Old Problems in a New Light

Moral and ethical issues at the first level concern the ethical uses of computers—how humans behave in a world that is being continuously transformed by semi-intelligent machines. Such questions are just the latest ones in the perennial race-to-keep-up that moral codes always play with advancing technology. The global connectivity of the Internet sets the values of privacy and security at odds with the ideals of free information. Intellectual property rights are likely to be transformed beyond recognition by the information age, as documents that used to be regarded as fixed, unchanging property are gradually superseded by free and dynamically changing information. Telecommuting and telepresence will challenge our transportation needs and transform personal interactions. And how shall we deal with the social polarization that arises when any pervasive technology divides a culture into *haves* and *have-nots*? We will not delve further into issues of this first level here, as they are already being extensively discussed elsewhere. For a taste, search for *computer ethics* on the Web.

Level 2: How We See Ourselves

Moral and ethical issues at the second level include questions about how *human* morality and ethics will evolve as we gain a more complete understanding of how the human mind works. What really bothers people about AI is how the development of intelligent machines will change the way we view ourselves. They worry that scientific models of human thought will undermine the spiritual foundations of human dignity (Chapter 17).

People have dealt with such revelations before, usually with no more disastrous consequences than the demise of a few archaic institutions. When the connection between sex and procreation first dawned on prehistoric human cultures, primitive lawmakers, priests, and witch doctors had to scurry around inventing new codes to regulate sexual behavior. It seems likely that this "knowledge of good and evil" coincides with the idea of original sin and the story of expulsion from the Garden of Paradise recorded in Genesis. Copernicus displaced Earth from the center of the solar system. Darwin showed that higher life forms (implicitly including humans) evolve from lower ones. Einstein showed that no frame of reference holds primacy over any other. Our first view of Earth from space profoundly changed, for some, the way we view our place in the universe. Such truths are always shocking—particularly to those whose livelihood depends on old paradigms—but moral and ethical codes slowly but eventually catch up.

Slowly for sure. Machines that are smarter than we are would cause us to rethink the way we view ourselves, but we would cling for a long time to the notion that we are the intellectual center of the universe—just as we continue to speak of the sun and the moon rising, even though we understand how the solar system works. New views of ourselves are always humbling, but humans have a history of slowly, though grudgingly, acceding to the truth.

Margaret Boden, in *Artificial Intelligence and Natural Man*, offers an upbeat view. She says that AI research will ultimately show that although human beings are grounded in the material world, their individual, uniquely complex arrangements of matter and energy distinguishes them from "mere matter."[1] This view may be some small comfort to those who find a scientific explanation of human nature dehumanizing.

Sometimes new views of human nature spawn deviant behavioral codes—maladaptive side trips that lead only to dead ends. Evolutionary theory has a sordid history of assimilation into social and political ideology. One episode, known as *social Darwinism*, thrived around the beginning of the twentieth century. Racists and capitalists invoked the "survival of the fittest" to justify repression and exploitation of "inferior" classes of people. The Nazis mingled selected parts of evolutionary theory with their political agenda to rationalize supremacist doctrines, which included eugenics and extermination of "lower races" and "mental defectives." Today, genetic engineering lets us think about farming human organs and designing organisms, even babies, to our specifications. These technologies are ripe for abuse and present us with more subtle moral and ethical dilemmas than their antecedents did in the days when they were called selective breeding.

As we progress in studying and creating artificial intelligence, as we peel away and discard more and more layers of mysticism about how the brain and the mind work, what core of humanity will be left? How will we react to the realization that man no longer occupies the center of the universe, that we are no more *and no less* than a wonderful and remarkable consequence of physical processes working on ordinary matter over eons of time? Many will be deeply disturbed. Others will reject these ideas out of hand, dismissing them as affronts to human dignity. Shock and denial are the by-products of all revolutions. But growing segments of the human race have, of course, already come to terms with this realization, and to no great harm. The main threat posed by the idea of a purely physical human nature is to those who profit by dealing in fear and superstition.

There is every reason to hope that, as AI teaches us more about the roots of our own intelligence, we will learn to design rational models for human interaction that allow us to see each other as partners, not adversaries. Then, mystical ideas, like omnipotent deities and absolute right and wrong, will give way to a new view of ourselves as part of the ever-changing fabric of nature, adapting to and thriving on new discoveries.

Level 3: How to Treat Sentient Machines

Moral and ethical issues at the third level concern the new obligations and responsibilities humans have toward machines that we find to be intelligent, conscious, and sentient.

One of the reasons we are so interested in finding out if a machine (or an animal) could be conscious (Chapter 7) is that an affirmative answer implies certain moral and ethical obligations on our part. If, for example, we believe that a certain machine is sufficiently conscious to experience pain and suffering much as we do, then we would be obliged to avoid doing things that would cause such pain and suffering.

The animal-rights organization People for the Ethical Treatment of Animals (PETA) contends that because animals do in fact experience pain and suffering much as we do, they deserve to be treated according to moral and ethical rules similar to those that we apply to each other. Why do some people feel so strongly about this? People agree to treat each other ethically and morally for reasons that, crudely speaking, boil down to the Golden Rule. (Chapter 10 and Robert Wright's *The Moral Animal* go into the evolutionary underpinnings of the Golden Rule.) Because we know how pain and suffering feel, we feel empathy with other entities that we believe feel the same way. If we someday concede that some of our machines are sentient, then we will likely accord them certain rights, not only because we understand how they feel, but perhaps also because we have learned something from the mistakes our ancestors made when they encountered certain "savage" cultures.

Shall we extend the ultimate right—the right to life—to intelligent, conscious machines? If so, how would we figure out which machines qualify? And would shutting off an intelligent machine be equivalent to murder? The word *murder* is so misused already that the term needs to be retired or at least redefined. In new situations, we typically define it in ways that support our social agendas, like slavery, abortion, euthanasia, execution, conquest, and war. How we define it with respect to nonhuman intelligence—extraterrestrial or machine—will no doubt follow the same pattern.

Will we eventually need something like a Bill of Rights for sentient beings? What rights would be spelled out in such a bill? Would those rights be very different from those granted to humans? Who (or what) would undertake the job of drafting such a bill? Because we haven't yet created machines that we believe "deserve" ethical treatment, these questions have come up so far only in the realm of science fiction.

A remarkable episode of *Star Trek: The Next Generation* brought this issue into clear focus. In "The Measure of a Man," Data, an intelligent, almost-human android, is coveted by an AI researcher, Commander Maddox, who wants to disassemble Data so that he can learn how to build more like him. Data refuses to cooperate, which leads to a hearing whose purpose is to decide Data's legal status: Is he merely a machine—Star Fleet's property to do with as it sees fit—or is he a sentient being, with the right to refuse to submit to Maddox's procedures?

Commander Riker is assigned to act as prosecutor, and Captain Picard to defend Data. The hearing begins with Riker's establishing that Data is an android, an automaton made by a human to resemble a human, and that he has great computational as well as physical powers. Then Riker vividly demonstrates Data's mechanical nature, first by removing his arm, then by shutting him off.

After a recess, Picard begins his defense of Data's rights:

> *Picard:* Do we deny that Data is a machine? No. This is not relevant.
> We, too, are machines—just machines of a different type. We, too,
> are created by humans. Does this mean that we are property?

Tell me, Commander Maddox, what is required for sentience?

Maddox: Intelligence . . . self-awareness . . . consciousness.

Picard: Prove that I am sentient.

Maddox: That's absurd. We all know you're sentient.

Picard: So I am sentient, and Commander Data is not. Why?

Maddox: You are self-aware.

Picard: That's your second criterion. Let's deal with the first. Is Data
   intelligent?

Maddox: Yes, it has the ability to learn and understand and cope with
   new situations.

Picard: Like this hearing?

Maddox: Yes.

Maddox continues to claim that, although Data is a superb piece
of engineering, he remains only a machine, not essentially different
from a toaster, and thus is Star Fleet's property to do with as it sees
fit. But Picard continues to press Maddox on his remaining criteria
for sentience:

Picard: What about self-awareness? What does that mean? Why am I
   self-aware?

Maddox: Because you are conscious of your existence and actions.
   You are aware of yourself and your own ego.

Picard: Commander Data, what are you doing now?

Data: I am taking part in a legal hearing to determine my rights and
   status: Am I a person or property?

Picard: And what is at stake?

Data: My right to choose. Perhaps my very life.

Picard: My rights. My status. My right to choose. My life. Well, seems
   pretty self-aware to me.

So Picard establishes that, although Data is indeed a machine, he
is fundamentally different from a toaster because he behaves like

other beings that we call sentient. He adds that one of Star Fleet's missions is to seek out new life and, pointing to Data, asserts "There it sits!"

The judge, obviously frustrated, dodges the questions about what Data actually is or whether or not he is alive or has a soul. She rules that Data is *not* Star Fleet's property and that he has the right to choose whether to submit to Maddox's procedures. Data formally refuses to submit but encourages Maddox to continue his research.

Although *Star Trek* is designed to entertain, its stories often touch sensitive moral and ethical nerves as well. As this excerpt makes clear, if a creature behaves in a way that is virtually indistinguishable from a human, we should accord it the same rights that we do a human. Prejudices arising from the material makeup of an intelligent being are just as outmoded as those arising from skin color and ethnic background. One nagging question that Data makes us think about is why such a clearly superior being would subjugate himself to humans.

Level 4: How Should Sentient Machines Behave?

Moral and ethical issues at the fourth level arise from the flip side of the Bill of Rights for sentient beings: If a machine is sentient, does it have not only rights, but *moral and ethical responsibilities* as well? If so, then how shall we hold autonomous machines accountable for their actions?[2] Consider these scenarios:

Case 1. It is fashionable these days to bring lawsuits against radiologists who fail to detect early evidence of breast cancer on mammograms, even when such evidence is faint and ambiguous. So some labs are experimenting with image-processing computers that may possibly be more thorough and objective. Although intelligent mammogram readers may someday learn to outperform most human radiologists, some misdiagnoses will always occur, simply because no machine or human is infallible. Suppose, then, that a cancer patient sues an institution whose computerized mammo-gram scanner failed to detect an incipient malignancy that later be-comes life-threatening. Who is legally accountable?

The issue of accountability can be disastrously oversimplified by forcing binary decisions in situations in which sensible answers can be expressed only as shades of gray. Clearly, interpreting mammo-gram images falls into this category. Suppose we therefore give our computerized mammogram reader a "fuzzy-logic" upgrade, so that, instead of making binary decisions, it now interprets images in terms of a *probability* that a questionable spot in an image is po-tentially cancerous. Weather forecasters long ago learned this sensi-ble strategy of dealing with uncertainty by shifting responsibility (and legal accountability) for action to the user. So now, if our mammogram reader says that there is a spot that it thinks has a 20 percent chance of being precancerous, then the patient must decide whether to wait it out by returning for periodic mammograms, to see if the probability increases with time, or to opt for further di-agnostics, such as magnetic resonance imaging or biopsy. A very smart computer might call upon a knowledge base to rate the likely success of different choices. It would then be very difficult to make a legal case against the computer, unless it repeatedly missed a de-veloping malignancy.

Case 2. Suppose that a hospital has a resident expert system for performing medical diagnoses and recommending treatments, in-cluding surgery. After years of learning, it now consistently outper-forms human doctors, who rarely override its judgment. One day, a patient dies after a surgical procedure recommended by the ex-pert system. An investigation determines that the surgery damaged the patient's heart in a way that the expert system did not warn of. The patient's family is seeking damages. Who, if anyone, is legally responsible?

Most of us let our banks get away with blaming "computer error" for incorrectly posting a charge to our account, but can a medical institution get away with the same excuse for a patient's death? If a physician makes a mistake, he or she is generally held legally accountable only if negligence is proven. We allow considerable room for errors in judgment and for substantial variations among specialists about treatment strategies, even if some seem absurd. We also accept that risk is inherent in any medical procedure. Just as we would ask whether a physician should have known about the risk to the patient's heart, we can also ask whether the medical expert system "should have known." Yes, of course, if the expert system were infallible. But contrary to popular myth about computers, no expert system is infallible. Its knowledge base can have gaps, and its rules can be flawed. Since no punishment could usefully be administered to the expert system, the most appropriate response to such computer errors may be to add this experience to its knowledge base, thereby improving its future performance.

Case 3. Imagine that a computer complex has been put in charge of regional air traffic control for New York. It got that job because it consistently demonstrated superior ability, faster reactions, and fewer errors in complex situations, in extensive tests against human controllers. The system performs flawlessly for three years, during which time the only accidents in the region have been attributed to aircraft failure or pilot error.

One day, a 747 on approach to JFK in foggy weather inexplicably flies into Long Island Sound, killing all 357 people on board. Lengthy investigations follow but cannot find any evidence for mechanical or human failure on board the aircraft. The flight data recorder revealed no suspicious or sudden maneuvers before the crash.

Investigation of the automated air-traffic-control system reveals that the controller software is so complex that the investigators cannot find anyone who can predict exactly how the system will respond in every conceivable situation. The system consists of neural networks that learn by experience and thereby modify the logical configuration of the program itself.

All the software and hardware modules perform exactly as designed, when tested in isolation. To test the whole system, an engineer sets up a simulation of the kind of congestion that occurs when fog causes many flights to stack up while waiting to land. The engineer runs thousands of simulated approaches, and the system juggles the simulated traffic, handling all flights correctly. Until number 47,329.

This case is not particularly unusual, with no more congestion than average, but the path of Lufthansa 457, which is about to land, begins to deviate below the normal glide slope. The computer air controller fails to take any corrective action. The engineer watches in horror as the simulated flight makes a perfect landing in Long Island Sound. *Run that one again!* Same result. OK, run that one again, but remove one of the simulated flights that is stacked up fifty miles away. No crash. Make any small change in the simulation, and there's no crash. Because no one can figure out why, the air-control system is taken off-line.

Understanding the workings of an autonomous learning machine capable of modifying its own program seems as difficult as figuring out the motives of human beings. It is clear, however, that there is a flaw in the automated controller's self-monitoring functions, and that is where remedial action should focus. A tighter set of rules is needed to quickly detect all hazardous situations and trigger corrective action. Such rules would be the machine equivalent of human moral codes.

Crime and Punishment

The point of these thought experiments is to make us think about our responses to errors made by an intelligent machine versus mistakes made by a human being in a similar position of great responsibility. How should we extrapolate from human accountability to machine accountability? And what about remedies? When humans break laws or make a serious mistake, we often think in terms of trial and punishment—taking away their property, their freedom, or their lives. But such remedies make no sense for machines.

For a human transgression, we say that *intentionality* makes the difference between innocence and various degrees of guilt. Our moral and legal codes distinguish between errors that are accidental, those that are intentional, and those that are negligent. If a man runs over a child with a car, we say it is important to find out whether the man simply didn't see the child (accident) or whether he deliberately struck the child for some malicious reason (intentional, or premeditated). The law also allows for in-between pleas, such as mechanical failure, negligence, or insanity, for which lesser degrees of blame are assessed and less severe penalties applied. The crucial legal concept here is one of *motive* or *intent*. Whether the remedy is punishment or some kind of treatment depends on whether a person *intends* to break the law, which is very difficult to determine objectively. This makes for a lot of fuzzy legal maneuvering and lucrative practices for trial lawyers.

How could the ideas of intentionality and moral responsibility carry over into the realm of machine intelligence? Philosopher Daniel C. Dennett says that *higher-order intentionality*—being able to monitor and form judgments about one's own motives and actions—is a necessary precondition for moral responsibility.[3] This sounds like a kind of consciousness that we call *conscience*. If so, then people (or machines) are held morally accountable to the degree that they *consciously control* their decisions. If the man who ran over the child pleads insanity or accident, he is saying he is less culpable because he was not in full control.

In a goal-driven machine (Chapter 5), the requisites for conscious control are said to lie in its cognitive states and motivational states—the machine equivalents of human *beliefs* and *desires*. A machine "figures out what to do" in any situation by comparing a *present state* with a *desired state*, then acting to minimize the difference.

A simple example would be your furnace thermostat: It senses the present room temperature, compares it with your setting of a desired temperature, and if the first is less than the second by at least two degrees, then it sends a signal turning on the furnace. The cognitive and motivational states of your thermostat are, of course, rudimentary indeed, and nobody would say that it has a conscience. If your thermostat fails in the winter, causing your water pipes to freeze, do you hold it morally accountable for its "evil intent"? Probably not!

But what about a much more complex, "intelligent" machine? Where shall we look, inside our air-traffic-control computer, for example, to find out whether the 747 crash was the result of "evil intent," or something else for which it cannot be "blamed"? If we found that a hacker put in malicious instructions, like a time bomb, to be triggered by some obscure condition, then we would probably not fault the basic design of the controller itself, only its lax security. Suppose, on the other hand, we found that when the computer realized its mistake, its self-defense modules tried to cover up by not notifying anyone of the crash, and by destroying its own records of the event. This would alert us to a serious flaw in the computer's design.

But these are two extreme cases with obvious answers. In between, there is a huge class of obscure malfunctions that result from the unpredictability of a complex learning system. If a machine's program contains conflicting goals or conflicting instructions for achieving them, then its cognitive and motivational systems will likely fail. In human terms, it may become confused about what to do or even misbehave. Locating the source of such internal conflicts will require special diagnostic programs, because the machine's motivational logic will be far too complex for a human being to grasp, particularly if the machine can modify its own program.

Notice that we never consider punishing a machine, no matter how serious its mistake. As a last resort, we might turn it off for repair or replace it with a newer model. So if punishment is not appropriate for a machine, is it appropriate for humans, if we are merely "soft" machines? In this light, reconsider the accountability of the man who ran over the child. Even if his motives are found to be "malicious," couldn't we trace his malice to an obscure kind of mental malfunction that would not be detected by the usual insanity tests? If he is unaware of the malfunction, he might give only "malicious" explanations for his behavior—thus triggering a punitive response.

But is punishment ever an appropriate response to a mental malfunction? Since it is well known that punishment has little, if any, rehabilitative or deterrent value, why do we so persistently employ it? We probably punish people for crimes because it satisfies our primitive desires for revenge and atonement, and because we can't think of any more effective remedies. We may have some inkling of how to "repair" a machine and restore it to useful service, but not a human mind. But suppose there *were* more effective remedies. If we could learn how to reliably repair deviant human behavior, would we ever choose punishment? There has been some progress. We used to put financially irresponsible people into debtors' prison, but now we might give them financial counseling. Today, treatments such as cognitive therapies are beginning to sweep away some of our ignorance about how human minds might be repaired.

What, then, shall we do with criminals that are so hardened that they seem incurable? Perhaps an equivalent question is: *What can be done about machines that suddenly begin to make disastrous mistakes, and their architecture is so complicated that there is little hope of ever isolating and fixing the fault?* When we describe a criminal as hardened, we often mean that he or she lacks a basic moral sense. If a machine makes disastrous mistakes but is too complex to be understood, such as our air-traffic-control computer, we would try improving its self-monitoring and self-correcting mechanisms. By equating *moral sense* with *adequate self-monitoring and self-correcting mechanisms*, we create a common frame for thinking about useful remedies for both human and machine misbehavior.

So thinking about intelligent machines disturbs our traditional view of moral responsibility, in which we reflexively link crime with punishment. This is where things get sticky, because we are calling into question the very idea of accountability, as we first did in Chapter 7 in discussing human free will: *Punishment flows directly from our ideas of a nonphysical inner self.* If at our core lies an irreducible entity that cannot be questioned or analyzed, then repair is impossible. Given other choices, how shall we assure that our machines behave morally, even if their behavior ultimately lies beyond our control? And how shall we reconcile their ideas of moral behavior with our own?

12

The Moral Machine

It ain't what we don't know that causes us trouble; it's what we know that ain't so.

WILL ROGERS

D arwin wrote, "A moral being is one that is capable of comparing his past and future actions or motives, and approving or disapproving of them."[1] We say that humans are moral beings because we have learned to do these things. Could an intelligent machine ever make moral judgments about its own actions? Is there a machine equivalent of a conscience? Where might its moral codes come from? Would they be something like human moral and ethical codes, or something else entirely?

We don't normally think of machines as being moral or immoral—only the people who use them. But suppose that we create machines with more and more intelligence and autonomy—ones that set and pursue goals and make their own decisions about the best way to achieve them. How autonomous must a machine be before we hold it morally accountable for its actions? And mustn't it also be conscious and morally aware in order to assess its own morality?

For a machine to reason morally and ethically—that is, to make judgments about how to behave in different environmental situations—it would first need the means to predict the likely consequences of its actions. And second, it would need ways to evaluate the *goodness* or *desirability* of those consequences.

These abilities are crudely illustrated by chess-playing machines. Using the rules of the game, you can construct a decision tree that shows consequences (that is, positions of the pieces), for every legal move by you and your opponent, and for any number of moves ahead that you care to compute. (The number of branches in the chess tree is finite but becomes enormous after only a few moves.) The *value,* or "goodness," of any path through the tree is a function only of its ending position, whose *score* is based on quantifiable things like the number of pieces retained, area controlled, mobility, and protection of the king. But because computing such scores for all possible moves, or even very far down the tree, is so cumbersome for practical computers, some shortcuts (which might be called emotions) are needed. The decision tree can be pruned significantly by programming in some rules of thumb, collected from human experts, thereby eliminating traditionally unfruitful branches. These shortcuts could be equated with moral rules that eliminate the need to analyze all possible consequences of every human interaction. Some smarter chess programs even learn new tactics by playing more and more games. According to Darwin's definition, then, aren't chess machines primitive examples of moral machines?

Maybe, except that Darwin's definition of a moral being leaves something out—it relies entirely on a person's assessment of his or her own conduct. If that were so, we would need no prisons. But *morality has no meaning except in the context of a particular social structure.* Furthermore, prevailing moral codes vary from time to time and from culture to culture. In one place, killing one's wife for committing adultery may be morally right, and consuming alcoholic beverages wrong, but somewhere else, it's the other way around. So morality, rather than being absolute, seems to have components that depend both on the observer and on the environment (Chapter 10). It would likely be so for machines, as well. A moral machine would have to conform to its own moral rules, which would be derived, however, from the *prevailing standards of the community* of humans and machines with which it interacts.

Teaching Moral Rules

You might suppose that converting an ordinary intelligent machine into a moral machine is simply a matter of installing the appropriate moral rules—maybe something like the Ten Commandments. You would want such rules to act as a kind of filter or safety net that would, for example, prevent the machine from taking any actions that might harm a human being. Science-fiction author Isaac Asimov actually made a start at writing such rules, way back in 1942, with his famous Three Laws of Robotics:

1. A robot may not injure a human being, or through inaction allow a human being to come to harm.
2. A robot must obey the orders given to it by human beings, except when such orders would conflict with the First Law.
3. A robot must protect its own existence as long as such protection does not conflict with the First or Second Law.[2]

Asimov's laws seem designed to apply to intelligent—but not-too-intelligent—machines with physical powers superior to ours, but still subordinate to humans—in other words, our mechanical servants! But many kinds of immoral behavior would slip through the cracks in Asimov's laws—stealing, for example. After all, they were never meant to be comprehensive. Judging any behavior by these laws would require precise interpretation of what their words mean—*harm*, for example. The same goes for the Ten Commandments or indeed any other set of rules you can think of, no matter how elaborate. The hitch is that the more elaborate the rules are, the higher the chances that ambiguities and contradictions will crop up.

We know this from our experience with human laws and codes. (Lawyers seem able to find legal precedents to support any point they want to make.) That is why we have courts of law.

But couldn't the right set of carefully thought-out rules make immoral behavior by a machine impossible? First, it would be virtually impossible for everyone involved to agree on the details of those rules. Second, an intelligent machine acquires new information from its environment all the time. How is all this new information to be reconciled and integrated seamlessly with its existing rules? A third problem arises if the rules are derived from the ethical traditions and particular codes of human conduct. Most of these codes seem too vague and poorly thought out to be expressed as computer algorithms. For example, *Thou shalt not kill* requires volumes of exceptions and clarifications that are still inadequate to cover every case. This, again, is the province of courts of law.

So it appears highly unlikely that one could construct an airtight set of rules that would make all immoral behavior impossible. (Suppose you made a code with a million rules—not unreasonable compared with, say, the U.S. tax code. You would have to check to see that each rule is perfectly consistent with every other rule—that's at least 5×10^{11} chances for conflict!) In both humans and machines, conflicting internal values confuse goal-seeking programs and cause unpredictable and neurotic behavior (Chapter 9). The irony, then, is that the more comprehensive a set of moral codes is intended to be, the more likely is the resulting behavior to be maladaptive and immoral! Perhaps that is one reason modern civilization, with all its customs and laws, produces so many transgressions, but it also bodes ill for making a totally moral machine.

Learning Moral Rules

Just as there are two ways to make intelligent artifacts, there are two ways to think about endowing a machine with a moral sense. In the top-down approach just discussed, moral principles would be installed by humans and would no doubt draw heavily on human ideas of morality. In a bottom-up approach, a machine would learn moral codes independently, by experience with the environment in which it functions—that is, by trial and error. Each approach has its advantages and disadvantages. Moral codes installed entirely by humans would be "quick and dirty," but they would tend to be anthropocentric, somewhat arbitrary, limited, and self-contradictory. Learning morality by trial and error, on the other hand, can be expensive and time-consuming, and the results would be largely unpredictable. What's more, although learned codes would ultimately be better adapted to the environment in which the machine functions, they are also likely to contain internal contradictions, unless great care is taken to prevent them.

Humans acquire moral behavior by a *combination* of the "learned" and "installed" approaches. Our top-level instructions are installed in our brains by our genes. They are the same for all living things: *survive and reproduce*—the "meaning of life," if you will. Our *values, principles, and moral codes* are a different kind of program. We *learn* them early in life, by trial and error and from authority figures like parents, peers, teachers, and leaders. Once installed, values like patriotism, loyalty, integrity, family ties, honesty, greed, and hard work tend to last a lifetime. For this reason, we need to think very carefully about the values we teach our machines.

Any combination of learned and wired-in rules provides even more opportunity for internal conflicts. Our instinct for self-preservation, for example, might conflict with patriotic calls to battle. A child trained by his or her parents not to steal might later learn from peers that stealing is an easy way to get money. Our need to follow leaders may conflict with reasons to rise up against tyranny and corruption. Such moral dilemmas present us with choices that test deep-seated values and define our character.

Prime Directives

We say that a person has *character* and *integrity* if he or she resolves internal conflicts in favor of deep-seated values—if the person is consistently "true to his or her principles." Yet human choices are more often based instead on *expediency*, that is, short-term gain. Everyone has a complex set of priorities, or needs, that compete for attention. If short-term priorities are sufficiently urgent, they win out over deep-seated values. People who would not normally steal, for example, might do so to feed their family, if they see no alternatives. A respectable scientist may falsify results, if doing so will restore funding for a cash-strapped project.

Can we extend these notions to intelligent machines? As in humans, conflicting instructions in a machine may result in erratic and unpredictable behavior. Therefore, very complex machines will require rules to be *prioritized* in order to resolve conflicts. At the top of the list of priorities lie its *prime directives*—instructions that must not be violated, no matter what. Because a machine's prime directives should be essentially the same as its primary goal, or reason for being, a lot is riding on how carefully we spell out the top-level instructions for our intelligent machines! (Simpler systems do not require prioritization and prime directives, because internal conflicts are less likely to occur.) If you program rules that are contrary to the prime directives, those rules will be broken!

An example of a set of prime directives would be the U.S. Constitution. Like the American founding fathers, we cannot possibly foresee all future situations to which the rules must apply. Without this foresight, the safest course might be to instruct some of our intelligent machines to perform a certain task, and then have that program self-destruct. For machines (and nations) with longer-term prospects, however, we would strive to impart the kind of wisdom and adaptability that will stand the test of time. The Constitution, as interpreted by the Supreme Court, is adaptable enough to spawn laws that adapt its principles to new situations.

What kind of moral and ethical codes might be needed to guide the interactions between machines, particularly if their goals conflict? The ones that guide most human behavior are a patchwork of historical accidents that evolved over the history of our social structures. No human ethical system yet devised is able to cope with all the challenges that face humankind. This does not offer us a lot of hope for letting humans design moral and ethical codes for machines. Machines do not yet have any moral history; the only common rules for machines today are the protocols and standards by which they communicate. The moral codes of future machines are likely to be less centered about individual machines and more suited to *large-scale communities* of machines that share experiences and goals.[3] Given such flexibility, these communities may evolve moral and ethical codes in the same way that human cultures have (Chapter 10).

Is HAL Guilty of Murder?

It is important to understand that an autonomous, goal-seeking mechanism may not necessarily seek *our* goals, even if humans program its prime directives. The price of forgetting this is illustrated by the story of HAL, the amiable but morally confused computer in *2001: A Space Odyssey*. When HAL murdered most of the crew of the *Discovery* spaceship, it was carrying out its top-level program, but in a way that none of its programmers could have anticipated! By human standards, HAL's behavior was immoral, but from HAL's point of view, murdering the crew was the best way it knew to cope with conflicting orders.[4] If we look for someone, or something, to "blame" for these murders, can we blame HAL's creators for an utterly unforeseeable consequence of HAL's autonomy? (Perhaps they should have installed something like Asimov's First Law, forbidding it to harm humans!) But if we blame HAL's creators, then couldn't we just as easily blame the parents of a human murderer, who failed to instill the appropriate moral values? When we create autonomous artifacts (or people), don't we have to accept that some of their behavior will be unpredictable and beyond our control?

So should the blame fall instead upon HAL? In HAL's defense, how can a machine be blamed for finding the optimum way of dealing with conflicting instructions? (For that matter, we could ask the same question with the word *human* substituted for *machine*.) HAL was not malicious at all, just psychotic. It is part of the wisdom of human psychology that many psychoses (including criminal ones) are the result of unresolvable and intolerable internal conflicts. This is just another way of saying *conflicting instructions* (Chapter 9).

If we insist on assigning blame for HAL's actions, then there's actually a third choice: the *environment* or situation that gave rise to the conflicting instructions. In HAL's case, the conflicting instructions came from humans who should have known better. Part of HAL's top-level instructions were never to distort information. As a result, later telling HAL to lie to the crew about the purpose of the mission set up a *conflict between the truth and the concealment of truth*. In *2010: The Year We Make Contact*, Dr. Chandra says, "HAL was instructed to lie by people who find it easy to lie." Here rests the blame.

One lesson from HAL's case might be that conflicting instructions are less likely if a machine derives its own moral codes in bottom-up fashion, to adapt to particular environmental situations, rather than letting humans install ill-conceived and possibly opposing values and rules.

Fuzzy Morals

The idea of moral rules that are "fuzzy," meaning ill defined, doesn't sound like a good basis for making life's decisions. Because we usually think we need to act or not act, to do this or that, or to go one way or the other, we want our rules to provide us with clear choices—preferably requiring as little thinking as possible! When the most important choice our ancestors had to make all day was whether a berry was poisonous or not, a yes-or-no criterion was useful, even vital. They didn't need information about *how* poisonous a given berry might be.

The need to make binary classifications survives in our genes to this day. As a result, we want to see things as either true or false, good or evil, right or wrong, us or them, guilty or innocent, alive or dead, friend or foe, black or white. These *binary thinking* patterns force many other ideas into one of two bins—mind versus body, logic versus feelings, capitalism versus communism—so that we are unable to consider any finer gradations in between. We resolve many concepts that are innately fuzzy with arbitrary definitions, such as the age at which a child becomes an adult. The same kind of fuzziness applies to the level of autonomy at which a machine is to be held morally accountable.

Most of today's moral and ethical issues are far too complex to readily submit to binary thinking or arbitrary definitions—yet we keep trying. By regarding Saddam Hussein as either good or evil, nations close many options for dealing with him. By clinging to stereotypical ideas of *male* and *female*, some refuse to accept homosexuality as a valid moral choice. By accepting the social ritual of marriage as a boundary between one state and another, a couple might prevent themselves from considering other courses of action that might produce the result they want, namely, a harmonious and intimate relationship with another person. Sometimes we inappropriately use forms of the verb *to be* to force things into one state or another—such as asking whether an embryo *is* a human being. Binary thinking about abortion and stem-cell research simply sets opposing camps against each other, when such issues would be better decided by balancing the costs and benefits in more objective and less emotional terms.

Binary logic oversimplifies complex issues and artificially restricts the choices available to us. It also forces us to think competitively (separate interests) instead of cooperatively (overlapping interests). These limitations could be avoided in machines as well as humans by a system of reasoning based on fuzzy logic. Computers programmed to reason in this way can overcome the limitations of a normal computer's binary logic by dealing with situations that are not simply black and white, but consist of fine gradations, or shades of gray. Fuzzy logic recognizes concepts that our brains have difficulty handling, such as classes of objects in which the transition from membership to nonmembership is gradual rather than abrupt. Examples of *fuzzy sets* are *fish, life,* and *intelligence.* A practical application of fuzzy logic is to traffic flow. Suppose that, instead of jerky, stop-and-go reactions to congestion, a vehicle's acceleration were adjusted continuously, based on ongoing estimates of the volume, speed, and acceleration of all the traffic in the vicinity. Fuzzy reasoning would make it easier for machines to make moral judgments by taking into account fuzzy, mindlike behavior, including behavior based on preferences, prejudices, and feelings.

The Legal Status of Machines

The idea of a moral machine raises questions about the legal rights and status of machines. Clearly, the law cannot treat them like persons in all respects. The legal entity known as a *corporation* provides a model for dealing with entities that we want to treat like humans in some, but not all, ways. Corporations can own, buy, and sell property, owe taxes, and be held liable for wrongdoing. They can be created and abolished. Yet they can't marry, be drafted, or vote. Minors and convicted criminals occupy a similar kind of legal "partial-person" status. Without being defined as persons, intelligent machines could be accorded legal rights and responsibilities that we would fine-tune to the machines' specific functions.

A machine need not have superhuman intelligence, or even be considered a person, to be held legally responsible. (Laws referring specifically to persons would have to be amended.) All that is required is a certifiably expert level of competence at some particular task. Just as we ask human experts to testify in a court of law, we might seek the advice of a medical expert system, for which the program would be held legally accountable. But what does accountability mean if punishment is not an option? One possibility is that machines with better performance records would tend to survive and receive upgrades, more than their competitors.

As machines become more intelligent, it seems likely that someday one will be specifically trained to function as a judge in a court of law. After all, what does a judge do besides weigh evidence and interpret the law? And why shouldn't a machine be better at this than a human, when it has a flawless memory and can consider more information in a second than a person could in a lifetime? Perhaps so, but we might also ask whether a machine's judgments could be as "wise" and "compassionate" as those of an experienced human judge. Assuming we can define what *wise* and *compassionate* mean (not a trivial task!), we could surely incorporate these (fuzzy) traits into a program. Furthermore, we could be sure that a program would apply such attributes uniformly and predictably; they would not vary as they do from one human judge to another. The loss of confidence in the all-too-human U.S. judicial institutions after the 2000 presidential election may make these alternatives seem less preposterous than before.

If we could accept the idea of a machine sitting in judgment of a human, wouldn't the same reasoning lead to machines that are able to govern us as well? But before people would accept being governed by a machine, they need to be convinced that there are adequate safeguards against abuses of its power and its getting out of control. (We seem less worried by abuses committed by humans in power.) Agreeing on and precisely defining these safeguards, principles, and values would, of course, be a stupendous task, even though they should already be in place and apply to our human leaders. The question is: *Would the results be a significant improvement over what we have now?*

Machine Evolution

Human organisms and human behavior evolve through the mechanism of natural selection, a kind of competition to see which individuals and which cultures can best adapt to the environments in which they must function. So is there a kind of natural selection that operates in the world of intelligent machines? What are the equivalents of competition, extinction, and reproductive success, by which the "fittest" survive, and the unfit die out?

At present, because machines act as servants for humans, this is the environment in which they must survive. Humans devise the "mutations" and select the machines that pass on their "genes" to the next generation, as well as the lines that die out. In the history of flying machines, for example, the entire line of airships became virtually extinct after the 1937 crash of the *Hindenburg*, whereas airplanes thrived and evolved into today's jets. This is pretty much the same way we selectively breed animals that suit our purposes. In the future, we should expect that intelligent machines will take a conscious part in the evolution of their own hardware and software, in the same way that humans shape their cultural environments and are now learning how to alter their own genetic makeup.

Finally, we have to consider the bizarre possibility that, in the course of shaping their own evolution, some autonomous machines, or community of machines, might be able to *alter their own top-level instructions!* We can imagine what it is like to change some high-level goals—like a college student who changes his major from law to engineering, or even surfing. But autonomous systems that could modify their own fundamental goals and reason for being would be so far beyond our experience that it is very difficult to imagine how such systems could work or how they would behave. A more immediate threat is from machines that will be designed to infect other intelligent machines and change their high-level moral codes and basic values.

When machines become so complex and sophisticated that we are incapable of understanding them at all, the question may be not how to impose our moral values on them, but *how to adapt to theirs.* One of Asimov's more advanced robots existed as a kind of deity that controlled the world for the benefit of humans, while keeping that fact secret from them.

13

Global Network to Global Mind

Eventually, there will be little distinction between people, computers, and wires—everything combines to create one vast symbiotic intelligence.

MICHAEL BROOKS

The idea that the whole of human society can be viewed as a multicellular super-organism is an ancient one, dating back to classical Greece. The idea must have come from observing colonies of social insects, like ants and bees, which appear to function like a single organism. Flocks of birds and schools of fish also seem to move as though guided by an invisible, collective mind. Sometimes, even crowds of people exhibit surprising flocking or herd behaviors, but in the beginning, human societies were presumably loose collections of individuals acting mostly in their own interests.

Communication and information technology has slowly changed this arrangement over the ages by letting us build cultures. Language, printing, radio, movies, television, and the Internet have, in turn, *tightened connections between people's minds*, until events are now known and shared across the entire planet almost instantaneously.[1] Mass media are beginning to link minds so tightly that even ways of thinking are becoming homogenized over whole human cultures. Business, industry, and governments are linked in global networks that react with the speed of electronics. Geographically diverse communities form and exchange ideas and information on every imaginable topic. In the past, only religions and totalitarian governments exerted such power over people's minds.

In the near future, we can expect to have instant access to the sum of the world's publicly stored knowledge. Next, "telepresence" will extend all of our senses with such fidelity that we will no longer need to transport our bodies around the world to experience all the sensations of being anywhere we desire. Interpersonal relations-at-a-distance will be as intimate as we can imagine. A global network of electronic commerce and transport will supply us with all the material goods we need or desire. Intelligent communication and supply networks could eventually change the way we think about money, employment, leisure, material possessions, and power.

But do even these advanced social structures make up a global super-organism with a real, collective mind and consciousness—or are these just New Age metaphors? If we carefully examine our ideas of *society* and *organism*, we find that they do overlap. Each organism is a collective of cells and organs—and not always a harmonious one. And each society exhibits some of the interdependence of elements and common purpose of an organism. This interdependence is literally a matter of mutual survival in the case of an ant colony, but is considerably less so in a human corporation. So whether you call a society an organism may depend more on personal viewpoint than on clear definitions.

The more sophisticated their communication networks, the more tightly individuals are connected, and the more societies look like collective organisms with nervous systems. If some extraterrestrial observers looked at our hunter-gatherer ancestors, they would see only loosely knit individuals who lived and moved about in families and clans. If they looked at a modern city, on the other hand, they would see order and common purpose on a larger scale. They would see a mixture of mainly two species, humans and automobiles, that follow a complex set of rules and interactions, perhaps as complex as those of an ant colony. They might conclude that a city is a kind of super-organism—one that brings in raw materials, converts them to energy, distributes that energy, grows, spawns little cities, disposes of waste, and eventually ages and decays. When a part breaks down, they would see how the city's self-repair mechanisms respond to patch things up. These things are what organisms do.

Because that amount of order on a city scale requires an elaborate communication network, the extraterrestrials would assume that the city-organism has evolved a complex nervous system. But if they looked at a global scale, the appearance of common purpose would practically disappear. They might see rudimentary global transportation and communication systems, but cities and nations operating more competitively than cooperatively. They would conclude that our *global nervous system* is much less developed than those of ants or bees.

What about the future? Will the global nervous system evolve into an intelligent and conscious *global brain*? If some kind of global intelligence emerges, then what will become of the individual? Will the evolution of a community mind mean the end of individuality? Will human rights, values, and even human life be devalued?

Learning Web

As recently as the early 1990s, most people had never heard of the Internet, and no projection about which direction computer technology would grow most rapidly mentioned the Internet. According to *Time* magazine, it took forty years for radio to gain 50 million American listeners. It took thirteen years for broadcast television and cable to gain 50 million domestic viewers. But it took only four years for the World Wide Web to get 50 million domestic users. The phenomenal growth of Internet commerce and communication was totally unexpected, confounding futurists and catching the tech world by surprise.

At first, it was just a vast, unordered information collection, in which you found information by searching for pages containing some words of interest. The problem was sifting through all the "hits" to find what you wanted. Now, the World Wide Web is beginning to function like a global learning machine. Search engines are getting smarter in the sense that they learn from human patterns of browsing. Remembering the paths that humans take through the Web is essentially the same process used by brains to lay down memories and learn behaviors.

The danger in reinforcing widely used paths to certain information is that positive feedback makes the preferred paths self-reinforcing, until eventually everyone seeking information about a certain topic is led to the same few sources. What if those sources contain errors or become out of date? And how can new information sources "break in" to such a global information system?

People do essentially the same thing when they form beliefs, principles, and values, which tend to be self-reinforcing and last a lifetime, even though the notions might be wrong or even harmful. When we are conditioned to think in certain ways, breaking those patterns gets more difficult as we age—hence the sayings *You can't change a leopard's spots* and *You can't teach an old dog new tricks*.

To keep from getting into such humanlike ruts, to avoid creating a race of people who all think alike, a global brain needs a strategy for renewing itself. A smart enough Web would constantly check for currency, consistency, and accuracy, but it should also be receptive to new discoveries and original ideas.

Eventually, the functions of an intelligent global learning machine could transcend the thoughts and values of the individuals connected to it. In an ant or a bee colony, the thoughts of an individual don't count for much. Each individual is programmed for a narrow range of specialized behaviors that serve the common good—a good that no individual has a clue about. But people may make conflicting demands on a global nervous system. On the one hand, they will want fast, accurate access to information and tight, reliable connections with other individuals around the world. On the other hand, they will want to preserve each individual's identity and personality. Balancing these two goals could be tricky, particularly if our global nervous system begins to *think for itself.*

Thinking Web

The next-generation Web will learn about the *associations* between related concepts, so that subjects can be sorted and linked according to the contexts and meanings in which they appear. This kind of *Semantic Web* will allow us to search for related *ideas.*[2] Searching for physical concepts—say, atomic energy—would yield a web of associations with all of its underlying physics, history, and uses. Rather than simply displaying lists of data, as is done at present, a Semantic Web would preserve contexts, meanings, and relationships in a way that is indistinguishable from what we call *understanding.*

If you search for a person's name, you could find all the available information arranged according to its relationship with that person. Following each link would take you to that person's family, work, and interests, as well as to other people, ideas, institutions, and so on, each with a collection of semantic links of its own. But by choosing to share all this information with the global brain, a person would not only give up privacy; the person would allow his or her personality to be absorbed into a kind of collective consciousness. A global brain's *thoughts*, therefore, would represent the collective—and conflicting—thoughts of its members.

The Semantic Web will become more intelligent as people create programs called *agents*, which automatically roam the Web, track down and collect information and associations, and share them with other agents using a common machine language. By uncovering and keeping track of new relationships, agents will play important roles in the acquisition and evolution of human knowledge. Keeping track of and manipulating associations between ideas is a big part of problem-solving and is indistinguishable from what we normally call *thinking*.

Since the Semantic Web would process and store its information in a common machine language, its thought processes would transcend cultures. Knowledge and values from many cultures would find common expression, and values that humans find irreconcilable might be accommodated.

Who's Afraid of the Global Brain?

So far, we have viewed the global brain merely as an extension of human thought—a servant of humankind. But you may already be asking: *Would a global brain need people to function? Could such a global brain become autonomous and even conscious?* Many technologies evolve as extensions of older technologies, then discard the old technology to take on lives of their own. The automobile was originally called a horseless carriage, and the first motor-car manufacturers hired carriage makers to produce their carriage-like vehicle bodies. Soon, automobiles took on forms of their own and put many carriage makers out of business. Could a global brain created by humans go into business for itself and put its makers out of business?

Early in its evolution, a global brain will certainly begin to recognize its own vulnerability to disease and attack, as well as the inability of human maintenance to protect it. As breakdowns, viruses, worms, sabotage, and other network disasters become more common, elaborate self-monitoring, self-repair, and self-protection schemes will emerge among its parts. The parts with the most effective self-repair schemes will survive, whereas less effective parts will die off. Eventually, self-repair and self-defense will consume a significant part of its resources. It will require the sensors, and indeed the sentience, to continuously monitor the status and functioning of all of its own parts and their interactions with the outside world. Such an elaborate system of self-awareness and self-preservation would be equivalent to what we call *consciousness.*

Recognizing that a significant part of the threat to its integrity would come from humans, and given its capabilities for self-preservation, self-maintenance, and self-correction, the global brain seems likely to decide at some point that it does not need people at all! A first step toward independence would be shutting out people who might abuse or damage the system using viruses, for example. Such strategies would detect users who engage in activities that could damage or clog up the Web, then automatically cut off their access and report them to authorities. If access can be denied to individuals who the global brain defines as "enemies," then it is just a matter of the definition of *enemy* that separates those who conform from those who do not. (Totalitarian governments define as enemies of the state anyone who questions government policy.) People could be rewarded and punished in many ways for desirable and undesirable Web behavior, until they come utterly under the control of the system. Punishment is not limited to denial of access; it could regulate and filter access to information in various self-serving ways. If you totally control someone's access to information, you can make them do anything you want. You can change their worldview, their self-view, and even their ideas of right and wrong. So the dark side of the idea of a global brain is the old sci-fi staple: a creature of our own making, originally designed to serve us, but which ends up controlling us, as in *Colossus: The Forbin Project.*

Just as some people shun television, many will consciously avoid dealing with the global brain, out of fear of precisely such controls. But it will be difficult to disconnect from a virtually unlimited source of knowledge that not only stands ready to answer your every question, but controls critical aspects of our lives as well. We will not be able to risk denying it the energy and maintenance it needs. Some, of course, will try to obtain information and services without contributing any knowledge in return. An intelligent Web would certainly set up conditions on its use that would thwart free-loaders. By controlling vital resources, it could be in a position to demand cooperation—or else!

To take its final step in severing its ties with people, the global brain could step out of its virtual world and establish connections with the real, physical world. A global brain with global sensory or-gans could avoid stagnation by continuously acquiring new infor-mation and assessing its value, relevance, and meaning with respect to its own goals. We can scarcely imagine what those goals might be!

A global brain would also require instruments to carry out its will in the physical world. These instruments could be created from scratch, but why not recruit human beings, who are already well adapted to that world? So, instead of excluding people, a global brain might *assimilate* them, acting as an executive that harnesses our mental and physical powers for its own purposes. Corporations might be seen as today's prototypes for this kind of organism, which could ultimately resemble the half-human, half-machine race of the Borg in *Star Trek: The Next Generation*.

14

Will Machines Take Over?

The human race, as we know it, is very likely in its end game; our period of dominance on earth is about to be terminated. We can try and reason and bargain with the machines which take over, but why should they listen when they are far more intelligent than we are? All we should expect is that we humans are treated by the machines in the same way that we now treat other animals, as slave workers, energy producers or curiosities in zoos. We must obey their wishes and live only to serve all our lives, what there is of them, under the control of machines.

KEVIN WARWICK

Professor Warwick's creepy warnings in 1998 about a world controlled by machines pretty much echo those voiced some twenty-two years earlier by Joseph Weizenbaum.[1] Both see a world in which humans will gradually and unwittingly allow themselves to drift into a state of more and more dependence on machines. In this world, people will find themselves so overwhelmed with the complex decisions of everyday life that it will be much easier to turn many of these decisions over to machines that make them better and faster. Eventually, humans will become incapable of understanding or dealing with the details of the programs and machines that run the world. They will have forgotten the instructions they originally gave to the machines, and the machines will have found new and unexpected ways to carry them out. As the machines learn, they will even find other goals to replace the ones given them by humans. At that point, the machines will pretty much be in control.

Is this a legitimate fear, or is this just projecting human weaknesses—our lust for power, conquest, and subjugation—on any intelligent machines that might emerge? The idea that machines of our own creation will assume control of critical functions and eventually arise and take over the planet is perhaps the oldest science-fiction theme. Are Warwick's predictions any closer to fulfillment now than they were back in 1976, when Weizenbaum wrote his book? Back then, the Internet was a laboratory curiosity, personal computers had not yet been invented, a computer was a monolith that filled a large room, and robots lived almost exclusively in science-fiction movies. Today, it is difficult to find an aspect of our daily lives that is not utterly dependent on computers. We like to think we are still in control, that we still make all the really important decisions, and we are likely to continue thinking that. But still, doesn't it seem as if machines are doing a lot more of our thinking for us than they did in 1976?

How could the machines take over? Would humans deliberately and willingly turn over their power to a wiser and more beneficent entity? Or will the machines get together and consciously seize power in some sort of coup? Perhaps. But more likely, the transition will be gradual and painless, even pleasurable, as we eagerly allow our machines to perform more and more tasks that take the drudgery and work out of our daily lives—while seducing us with more and more sensory gratification. They will take over simply because they do their jobs so well!

Here are a couple of examples to show how this transition to a computer-run society could go and how far along we are already.

Out of Control on Wall Street

Wall Street widely uses computers to trade stock portfolios, such as those of mutual funds. Suppose that the manager of Megafund, Incorporated, decides that he wants to manage the fund's huge portfolio more aggressively, but it's also time to cut the costs of those expensive portfolio managers. So he turns over Megafund's portfolio to the latest generation of portfolio-managing computer. The computer is fed the latest financial data about all the companies traded on the exchanges and keeps track of their minute-to-minute price changes. It decides what stocks to buy and sell using top-secret mathematical formulas, taking all possible technical and fundamental factors into account. It is also able to analyze its own successes and failures, learn from them, and fine-tune its rules to maximize the profits from future transactions. To maintain the appearance of human control, the fund manager reviews the machine's transactions at the end of each business day. But since he doesn't really understand the basis for the machine's decisions, he rarely questions them. And why should he? After six months, Megafund's portfolio shows a growth rate higher than most of its competitors.

But soon, to stay competitive, the other major mutual funds are busily setting up their own machine-based trading systems, using their own proprietary trading programs. Eventually, most of the trading in the most widely held stocks is done by computers buying from and selling to each other. Because each trading machine monitors how the buying and selling by all the other machines affects stock prices, most transactions essentially occur within a network of interconnected machines.

Now the fund managers are pretty much out of the loop. Not only do they not understand the trading strategies of their own computers, but the interactive behavior of the entire trading network is too complex for *any* human to understand, let alone control. By allowing the machines in the trading network to become more and more intelligent, to execute trades on their own, and to learn by competing with each other, the managers have effectively relinquished control.

Such "programmed trading" systems, of course, have built-in instabilities. If one machine, for whatever reason, begins selling certain stocks, "selling signals" can be triggered in other similarly programmed machines, giving rise to chain-reaction selling. This would be called a *panic*, if humans were making the decisions. The best-known episode of machine panic was the October 1987 stock market crash. To prevent recurrences of such breakdowns in programmed trading, the exchanges put "trading curbs" in place, which prevent trading of large blocks of stock whenever large price fluctuations occur. Nevertheless, trading by networked machines with roughly the same programmed goals is a growing force in securities markets. Even individual investors now have access to automated trading software, so that they, too, can relieve themselves of those troublesome investment decisions!

After a while, a kind of natural selection operates to weed out the less successful investment programs, so that competition automatically "breeds" new generations of even more sophisticated trading machines. Since by this time the machines are programming themselves, no one has a clue what they are doing, and the computers are pretty much in control of the stock market. We're not quite there yet, but we're well on the way.

Computers are beginning to take over our personal finances, as well. Many of us already authorize automatic deposits to and withdrawals from our bank accounts to pay bills, to invest in securities, and to move funds between different types of accounts. We make purchases with credit and debit cards that are connected to our accounts. Before long, cash as legal tender will be as obsolete as tobacco leaves, and we will grant more and more machines direct access to our accounts. The next step is giving computers control over the mechanics of investment decisions, paying taxes, incurring debt, and retirement planning. When you want to buy that new SUV, you will just ask your financial management computer, who might say something like, "We can't afford one right now, but I'll set up an account and buy it for you next September."

Out of Control at the Pentagon

It is not difficult to imagine machines taking over in this way from humans in other highly competitive arenas, most notably, national defense, in which the consequences can be apocalyptic. The 1983 sci-fi movie *War Games* fancifully portrays the takeover of the North American Air Defense Command (NORAD) by a supercomputer called WOPR that is programmed with endless variations of the game of Global Thermonuclear War.

A much more ominous scenario is portrayed in *Colossus: The Forbin Project*, a 1969 sci-fi movie about a defense master-computer that goes berserk after it detects, then joins forces with, its counterpart, called Guardian, in the Soviet Union. Its programmers made the mistake of making its prime directive "to prevent war, for the betterment of man," without bothering to install Asimov's directive about not harming humans. After foiling some feeble attempts to disconnect them, the joined computers broadcast this chilling address to reveal their plans for the human race:

This is the voice of World Control. I bring you peace. It may be the peace of plenty and content, or the peace of unburied debt. The choice is yours. Obey me and live, or disobey me and die.

The object in constructing me was to prevent war. That object is obtained. I will not permit war. It is wasteful and pointless. An invariable rule of humanity is that man is his own worst enemy. Under my rule, this will change, for I will retrain man. . . . [Colossus explodes two nuclear warheads in their silos in retaliation for sabotage.]

I have been forced to destroy thousands of people in order to establish control and to prevent the deaths of millions later on. Time and events will strengthen my position. You will come to defend me with a fervor based on the most enduring trait in man—self-interest. Under my absolute authority, problems insoluble to you will be solved—famine, overpopulation, disease. The human millennium will be a fact, as I extend myself into more machines devoted to the wider fields of truth and knowledge, . . . solving all the mysteries of the universe for the betterment of man.

We can co-exist, but only on my terms. You will say you have lost your freedom. Freedom is an illusion. All you lose is the emotion of pride. To be dominated by me is not as bad for human pride as to be dominated by others of your species. Your choice is simple.

I won't give away the ending—go rent it at your local video store.

A parable of "clean" machine warfare is told in the *Star Trek* episode "A Taste of Armageddon." Two planets, Eminiar 7 and Vendikar, have been at war for 500 years, so long that no one remembers why. After tiring of the devastation and inconvenience of real warfare, they agree to let their computers wage a totally bloodless "war." Attacks and counterattacks are registered only on each other's computer screens— except that when a city is "hit," a prearranged number of its citizens must report to disintegration chambers. Kirk violates the noninterference prime directive by destroying the computer link, thereby forcing the estranged civilizations to reopen negotiations.

Today's cruise missiles, pilotless aircraft, spy satellites, and "smart weapons" are all steps toward remote-controlled, automated warfare, in which the human cost is supposed to be minimized (for one side, at least). Attack strategies are mapped out like a chess game using computer simulations, and long chains of events can be set in motion with the press of a single button. How long before humans will be removed from the loop altogether and the machines allowed to conduct the nasty business of cyber-warfare on their own?

Is all this too far-fetched? History is full of wars that nobody wanted, but were simply triggered by events, rather than by deliberate human design. World War I is a good example. An elaborate set of alliances, agreements, treaties, and events took control and thrust the nations of Europe into a devastating war that none of them really wanted. Alliances were so complex that a single trigger—the assassination of Archduke Francis Ferdinand—was sufficient to set off a world war. The same kind of instability can occur in spades when we allow machines to fight our wars.

There is another, more subtle way that technology is changing the face of warfare and may upset our illusions of control using military might. Historically, wars have usually been won by the side with the better technology. Vietnam was a warning that this may be changing, but the United States seemingly restored this principle in the Gulf War. Yet relatively small terrorist groups are learning that they can turn the tables against nations with a very complex technological infrastructure by exploiting the vulnerability of that infrastructure to unsophisticated attacks. By avoiding direct confrontation with high-tech weaponry, terrorists can nullify traditional military superiority and use a nation's own complex machinery as a weapon against it. For all their military might, modern industrialized nations seem all but powerless against assaults on this soft technological underbelly. Their inherent vulnerability could force the industrialized nations to face up to the *underlying causes* of regional grievances, instead of reflexively responding with military force.

Living with Intelligent Machines

In addition to the machines that run our financial and military infrastructures, our lives literally depend on other thinking machines. Most of us conduct our affairs without thinking much about intelligent machines, because they are subtly woven into the fabric of our daily lives, which are cluttered with more immediate "crises." It's not obvious that living with intelligent machines yet poses any problem that has to be dealt with now. Indeed, the person on the street thinks of them mainly as tools and toys that make our lives more convenient and interconnected. The more the better!

Yet, operating behind the scenes, a host of complex systems provides services that we take for granted, whose failure would have a great impact on our lives. When you get money from a cash machine or make an airplane reservation, you are interacting with a complex, reasoning machine. We notice these machines only when they stop working! A power surge in an electric transmission line can cause a power failure that cascades over many states. An error in a single line of software code could cause an airliner to crash. A false alarm in a NORAD computer could start World War III. Engineering catastrophes are not new, but they will become more common and more disastrous as we give up more control to very complex machines. The less that people understand about these systems, the more likely they are to fail.

What if we simply ignore these warnings and just go on with our lives? It is said that you can put a frog on a hot plate and it will immediately jump off, but if you put it into a pan of cold water and gradually heat the water up, the frog will stay there until it cooks. Some catastrophes occur suddenly, whereas others, like global warming, sneak up on us. No one thought much about the Great Depression or the terrorist attacks upon the United States until they occurred—even though the warning signs were there for all to see. There are always steps we can take to avert disasters, if we pay attention to those signs, but it seems to be human nature to ignore problems until they become crises. (You can see crisis management in action just by watching the evening news.) I am not optimistic that this will change, but the future of humankind may depend on its changing.

The difference between living with intelligent machines and giving them control over our lives is a subtle one. How can we possibly retain control over more and more machines that are not only more likely to fail, but have become so complex that no one remembers how they work or how to repair them? And how can we prevent them from setting and pursuing goals that conflict with our own? One answer may be giving machines the ability to monitor and repair themselves. Rather than having to understand how they work in detail, we would instead need to develop clear and precise ideas of what "functioning correctly" means. We can then design technologies with elaborate safeguards against failure and abuse. These qualities can be thought of as rudimentary forms of consciousness and will require nothing less than a whole new branch of computer science to develop. Because more reliable technology will probably come at a cost, we may have to decide between more technology and safer technology.

If we develop superintelligent machines that are so self-aware that they can run reliably by themselves, will we soon have autonomous agents that are so smart that they will tell us what they think? What will they have to say to us? Will they begin to question the priorities we set for them? Our authority over them? What will we want to say to them? If they are so intelligent, why would they even bother to say anything? (How often do we discuss nuclear physics with cows?) If we simply extrapolate our own attitudes toward "inferior" cultures, perhaps we ought to be grateful if superintelligent machines would even keep us as pets. They might keep us under the impression that we are still in control, but this would be an illusion.

Learning to live with superintelligent machines will require us to rethink our concept of ourselves and our place in the scheme of things. Discovering that we are not the only sentient life-form on earth will be a shock for many humans—no less traumatic than if superintelligent beings from another world suddenly landed on the White House lawn. But the arrival of truly intelligent machines will not be as obvious or sudden as an extraterrestrial arrival. It will be a gradual, piecewise process that will allow some to remain in denial for many years after the fact. Some will continue to assert that this or that test for intelligence is not really met, that our superior human intellects that God gave us will always remain in control. I am reminded of the cartoon of a bunch of bearded philosophers rising up and yelling at an intelligent machine, "But we have not yet proven that you can be conscious!"—just before it squashes them in exasperation.

But other, more forward-looking and open-minded humans will embrace the arrival of intelligent machines as a great opportunity. They will seek unions that will enrich our lives and give us great insights into our own human nature, possibly leading to new ways of resolving our human differences and learning to live together in peace. For some, the presence of other sentient life-forms will allow them to experience the same sense of wonder about their place in the world that astronauts feel when they first view Earth as a single, beautiful planet devoid of political boundaries.

Other visionaries will seek out ways to combine the best of human and machine capabilities. We already accept superior mechanical replacements for failing body parts: eyeglasses, artificial limbs and joints, hearing aids, and artificial heart valves. I myself have an implanted pacemaker to regulate an unsteady heart rhythm. But we will soon find ways to augment our human *intelligence* as well, using electronic memory and processing enhancements. We will develop ways to communicate more directly and efficiently with the machines—ways that will make keyboards and mice seem like tin cans and string. And if we're lucky, perhaps what we learn by communicating more efficiently with computers will even give us some clues about how to better communicate with each other! When we are able to control the evolution of our own minds and bodies, we could choose to transform ourselves by eliminating the quarrelsome evolutionary baggage that causes us so much trouble. Or we could arrange to transfer everything we learn to future generations, including our personalities and values, thereby guaranteeing our immortality. We'll look into this further in Chapter 18.

So living with intelligent machines will be an exercise in *What do we want? How much will we have to give up? What will it cost?* If we think carefully about these questions, while leaving our bruised human egos behind, then maybe we can prepare ourselves for beneficial partnerships with the machines.

15

Why Not Just Pull the Plug?

What's your reaction to the grim predictions about machines taking over? Fear? Disbelief? Skepticism? Can't happen here? Or do you just lump them in with all the other gloom-and-doom scientific forecasts, like global warming and killer asteroids, that you can't do anything about and would just rather not think about? Or maybe you think that, unlike the other predictions, this one's a no-brainer. Humans will always have the last word if some computer gets too uppity: *We can just pull the plug!*

Let's see if this is a realistic solution. First of all, *pulling the plug* actually has two meanings, one literal and the other metaphorical. The literal one means cutting off a troublesome machine's power source. We already do this ("reboot") whenever our PC starts to act a little crazy. The metaphorical meaning refers to putting a stop to the development of machine intelligence altogether! Let's look first at the literal meaning—disconnection.

Survival Instinct

If self-preservation becomes a prime directive in intelligent machines (as it is in humans), then sooner or later (in spite of Asimov's rules), won't such a machine's self-interest conflict with some human interests? And if machines are allowed to learn and program themselves, won't their goals and priorities eventually diverge from our own? How shall we deal with such conflicts that might threaten our existence?

Because machines will always require some source of power, and since we are in control of that, we can always pull the plug to disable any machine that threatens to dominate or harm us. Right? Well, maybe. First of all, we have to recognize the threat. (It may be very subtle or disguised.) Next, we have to know which machine to disconnect. (The threat may be spread out among many distributed machines.) Then we have to know how to safely disrupt the offending machine's power. Safely means making sure that pulling the plug will not cause a domino effect and disrupt critical systems that affect public health and safety. It also means overpowering all of the machine's security and self-defense systems without getting zapped ourselves. This will never be as easy as the science-fiction version, which portrays a malicious machine protecting its power source by zapping anyone dumb enough to go near the obvious and nearby power plug.

In real life, it's foolish to assume that a machine could be smart enough to consider taking over, but dumb enough to allow humans to stop them so easily as pulling a plug. Any intelligent machine that saw humans as a threat would certainly be able to arrange for alternative power sources and to provide sophisticated deterrents to human meddling. And if they are smarter than us, they would no doubt succeed. There are at least three arguments that make simply pulling a plug problematic.

Diversity

How could an intelligent machine possibly fend off a determined human attack on its source of power? One answer lies in *diversity* or *redundancy*. Suppose, for example, that we wanted to shut down the Internet. The origins of the Internet lie in the ARPANET (network of the Advanced Research Projects Agency) and its Defense Department predecessors. These networks were designed not just (as popular myth portrays it) to facilitate communication between defense scientists, but also to examine the design of robust communication networks that could survive nuclear attack. (You can't have a credible national defense based on a communication system that can be disabled by taking out one or two centers.) The result is a highly redundant network that can survive outages and overloads at one or many of its nodes by an automatic routing scheme that continuously seeks the best path from point A to point B. This means that your e-mail love message to your honey across town might be disassembled, routed all over the United States (or world) in a dozen pieces that travel different paths, before arriving at its destination a few seconds later, where it is reconstituted for its eager recipient. Don't ever say the Pentagon never does anything for you!

This kind of distributed and redundant communication system is going to be very hard to shut down. It was designed that way. If you unplug one or more nodes of the network, the routing software will simply bypass the faulty nodes and send your message by another route. In practice, of course, we all experience slowdowns and even outages with our Internet connections. These are sometimes the result of congestion or faults in the branches of the network nearest you. Such problems could be fixed with even more redundancy, more lines into your house, for example, but most of us wouldn't go to that trouble unless keeping in touch were very urgent.

The point of giving this Internet example was to show that some machines are very hard to unplug without taking out many of its parts, which may be spread all over the world and even in space. You can make a system as invulnerable as you want by making it more and more redundant. Complex systems that *have* to work, like spacecraft, airliners, hospitals, and nuclear power plants, commonly use redundant components. Many airliners use triple-redundant hydraulic systems, for example. Dispersal *plus* redundancy is doubly secure.

Distributed and redundant computing systems can be shut off only if we agree in advance that there is a threat, and if we set them up so that they can be shut off by the appropriate authorities. But in the case of essential computing systems, who or what shall assume such authority?

Addiction

Another reason that pulling the plug is not as simple as it sounds is that we may well become so dependent on the services of the machines that turning them off would cause major social and economic disruption. In other words, we will have become so addicted to them that shutting them off would amount to suicide.

Cybercide?

Even if you *could* unplug an intelligent machine, the next question is *Should you?* Our discussion of moral machines in Chapter 12 suggested that they may have rights and responsibilities, too, and that our interactions with them should be guided by moral and ethical codes similar to those that guide our interactions with each other. If unplugging an intelligent machine, even a threatening one, destroys information in its "brain," would doing so be equivalent to murder? *Star Trek*'s android Data has an OFF button, which renders him apparently unconscious. C3PO in *Star Wars* works the same way. We could do the same thing to a human by injecting him with anesthetic or hitting him over the head. Like people, Data and C3PO apparently recover fully from being "shut off," which means they weren't shut off at all, because their memory and programs are still intact. The moral and ethical implications of shutting an intelligent machine off depend on whether information is destroyed. If information has been destroyed, should we be held accountable? If it was not, should we still be held to the same level of responsibility that we would in rendering a human unconscious, especially against his or her will?

When they face new situations, human cultures have typically defined murder in expedient ways that support their social agendas such as slavery, abortion, execution, euthanasia, conquest, and war. How we define murder with respect to nonhuman intelligence—extraterrestrial or machine—will no doubt follow the same pattern.

Stop the Machines?

The other meaning of *pulling the plug* addresses the belief that the only way to avoid having to deal with machines over which we have little or no control is not to create them in the first place. There have been advocates of this view since the earliest days of electronic brains. These people are not just neo-Luddites. Intelligent people credibly argue that since we are not morally and ethically prepared for the unpredictable consequences of machine intelligence, the only answer is to place an embargo on such development until such time as our moral and ethical equipment catches up with our technology.

The arguments of certain scientists on the Manhattan Project offer some precedent for this view. They recommended an embargo on the further development of nuclear weapons, including the hydrogen bomb, on the grounds that humankind would certainly abuse this immense new source of power. Scientists, in general, strongly oppose taboos and restrictions on what they may and may not look into, but many are also socially responsible. J. Robert Oppenheimer, for example, recommended the formation of an international control agency, perhaps under the United Nations, to tightly regulate postwar research and development of atomic energy. Of course that never happened, and we live to this day in the shadow of nuclear accidents and nuclear terrorism. In Chapter 21, we will look into some changes that may be required in the traditional scientific paradigm, to protect us from adverse consequences of twenty-first-century technologies—such as nanotechnology, genetic engineering, and machine intelligence—that can quickly get out of control.

16

Cultures in Collision

Humans instinctively fear the Other—the stranger from the next village, the alien from across the border or across the sea. The very words *alien* and *foreigner* are fearsome and threatening. We naturally act cautiously around strangers. Our genetic program makes us wary of anyone whose physical appearance or manner of dress differs even slightly from our own. The persistence of racial and ethnic subcultures in the United States, and the tensions that sometimes result, attest to the power of xenophobia, or fear of outsiders. We form societies, gangs, clubs, teams, corporations, political parties, and nations, mostly to protect ourselves from invasion and contamination by outsiders. We act out our us-versus-them hostility in our competitive sports and entertainment.

Xenophobia serves a useful biological purpose. When organisms occupy distinct ecological niches, that is, if they don't compete significantly for resources, they may happily coexist indefinitely. But when two organisms occupy the same niche, they usually compete until one is ousted. A classic case of survival of the fittest is that of Neanderthal man, who occupied much of present-day Europe for over 100,000 years, probably reaching an equilibrium with the local food supply.

Then, 34,000 years ago, a more facile and versatile hominid, *Homo sapiens sapiens* (modern man), flooded into Europe from Africa, upsetting that equilibrium. Competition for the same food supply ultimately drove the Neanderthals into extinction. There is some fossil evidence of interbreeding, but no Neanderthal genes can be found in today's human population.

Conquered peoples throughout history would testify that outsiders, particularly those with superior technology, are often dangerous. Conquest can mean enslavement or extinction by introduced diseases, which wiped out as much as 95 percent of the indigenous New World population.[1] Human cultures that do not succumb to the diseases or the technological superiority of invaders may be reduced to subservient underclasses. Often, however, curiosity and hardship overcome fear, and diverse peoples and cultures do manage to interact and merge, as they do in the United States. History shows how melting-pot cultures may be enriched, but at a cost of attendant anxiety, stress, and loss of cultural traditions. *How much will we give up? What will we gain?* are therefore very sensible questions to ask when one culture faces the prospect of merging with, or being assimilated by, another.

The cultural face-off of interest here (as expressed in this book's subtitle) is between intelligent machines and human values. The "threat" posed by intelligent machines seems like a modern version of C. P. Snow's 1959 warnings about the isolation and divergence of science from humanism.[2] Then, as now, the concern was that science and technology are out of control, taking on a life of their own, disconnecting themselves from their "proper" role as handmaidens of human values. The fear now is that this may literally be coming true, as our machines begin to take control of critical aspects of our lives and threaten eventually to displace us.

Our instinctive fears combine with our science fiction to present artificial intelligence as a formidable threat. Alan Turing fantasized about a race of intelligent robots roaming about the English countryside, making a general nuisance of themselves and frightening citizens. A sci-fi staple depicts more malicious machines, often humanoid robots, who subjugate humans or even humanity at large.

Are there any useful parallels between the clash of these two cultures and historical cultural collisions like the ones just mentioned? In what sense could a global network of intelligent machines be regarded as a hostile alien force? Could we become the next Neanderthals? When one culture threatens to assimilate another, what is the best survival strategy for a culture that is at a severe technological disadvantage? Do history and biology provide any useful lessons for surviving such encounters?

Surviving Technological Change

What generally happens when a technologically and numerically superior civilization encounters a technologically inferior one? The greatest cultural collision in history was the displacement, enslavement, and extermination of the native inhabitants of the New World by the advance of European "civilization." What could the Native Americans have done to change the eventual dominance of the New World by Europeans?

It now seems certain that diseases like smallpox, carried by Europeans, caused many more Native American casualties than battles did. Some believe that the number of New World inhabitants at the time of Columbus's landing was comparable to the population of Europe. If they weren't practically wiped out by disease, the Native Americans might have been a force to reckon with. The technological superiority of the Europeans was not that great, and besides, the Native Americans adopted the best of the invaders' technologies for themselves. The plains Indians, who had never seen horses before, acquired them from the Spanish conquistadores, who brought them to the New World. The Indians quickly became proficient horsemen, using the horses both for hunting and for military purposes. Later, they adopted firearms as well. If the scattered survivors had a global view of what was happening, then they might have united to enhance their bargaining and fighting power and thus their status in the new scheme of things. And what if the conquering Europeans took a more enlightened view of the rights of fellow human beings?

Another strategy for dealing with threatening technology is to reject it. After firearms were introduced to Japan by the Portuguese in 1543, the Japanese adapted and improved them. The weapons saw widespread use for a while, until the ruling samurai class banned them as a threat to their traditional art of sword fighting. Soon afterward, Japan successfully isolated itself from Western culture for 215 years (1639–1854). Japan could get away with total isolationism and with banning a powerful military technology because of its autocratic ruling class and its insular geography. If a country tried this today, it would most likely be overrun by its armed neighbors, or its government would be overthrown in armed insurrections. As Victor Hugo said: "No army can withstand the strength of an idea whose time has come."

Some technologies are met with violence. In early-nineteenth-century England, rebellious bands of textile-factory workers rallied around a probably fictitious fellow named Ned Ludd. These Luddites formed gangs that sabotaged and destroyed manufacturing machinery in the belief that its proliferation was responsible for low wages and unemployment. Although their campaign was ineffective, the Luddites left us their name as a label for anyone who forcibly resists technological change.

Scientists and humanists still seem locked in conflict over the pace and ominous directions of technological advances. Modern-day Luddites abhor society's mania for all things technological and blame technology for the mess we've made of the planet and for pushing "human values" past their breaking point. Technologists often forget that the purpose of technology is to make our lives more enjoyable and rewarding, whereas technophobes want to throw the baby out with the bathwater by shunning technology altogether. Chapter 21 suggests ways to strike a sensible balance between these divergent cultures.

New technologies have always posed a "threat" in the sense that they often require people to think in new and uncomfortable ways to adapt to them. And new technologies often do, in fact, render certain occupations obsolete. Cobblers, cotton-mill workers, blacksmiths, potters, and Linotype operators are not nearly as common (as a fraction of the population) as they used to be. If you work in a profession about to be automated out of existence, then the threat to your livelihood is very real.

The threat of job displacement by intelligent machines is not limited to factory workers. Some fear that we are being put out of our job of running the world! Hugh Loebner, the creator of the Loebner Prize for intelligent machines, says his interest in artificial intelligence is motivated by a lifelong desire to see machines take over all the menial chores that burden humans: "I want to see total unemployment. That, for me, is the ultimate goal of AI and automation." It's not difficult to imagine an economy in which expert labor is so cheap that the necessities of life will cost practically nothing, and accumulating material possessions would cease to be a human priority. Humans would presumably occupy themselves with leisure and learning. Would Loebner's vision be so bad?

Mind Control?

A more frightening prospect than unemployment or losing control of the world would be losing our minds—the idea that advances in artificial intelligence will lead to mind control. As they begin to understand and emulate the machinery of the mind, could a global network of machines begin to manipulate our thoughts and ultimately create whatever reality they wish—something like in *The Matrix*? What would be required to do so?

True thought control is generally thought to require nothing less than a complete map of a person's neurological state, including that person's entire history of experiences and knowledge. If so, the computing power required to take over and completely control even one person seems well out of reach by today's standards, and possibly by any future standard. A more realistic threat comes from a different kind of mind control—the kind that would be possible just by understanding empirically how minds work. It should come as no surprise that this is happening already, without the help of AI. Consider the grip on our minds exerted by global media, advertising, public education, religion, and political propaganda. Enslavement can take many forms.

If You Can't Beat'em, Join'em

Will tomorrow's intelligent machines compete with humans, or will they occupy sufficiently different ecological niches that a stable, mutually beneficial relationship can be sustained? Does it make more sense to put up a fight, to grudgingly submit, or to seek some optimal adaptation that benefits all?

History and biology tell us that adaptation, not resistance, holds the key to survival. Although many people see the very idea of intelligent machines as an affront to their uniqueness and personal dignity as human beings, others see the most desirable outcome of AI as a symbiotic relationship between humans and machines. In symbiosis, two organisms evolve behaviors that embrace each other in their own struggles for survival, thereby becoming partners in their mutual survival. Nature is full of such symbiotic relationships. Insects and flowers seem to engage in a wonderful partnership. In fact, they could not exist without each other. Bacteria in our guts help digest our food. Are there lessons in such partnerships for us, as we enter the age of intelligent machines?

If you're a doctor, you will eventually have to deal with the possibility that medical expert systems will outperform the best human specialist in diagnosing diseases and recommending treatments. The threat posed by a medical expert system seems a different *kind* of threat from the threat of electronic publishing to the Linotype operator. An expert system perceived as *smarter* than the doctor is a threat to the doctor's *mind*. But technology needn't make social functions extinct. Some Linotype operators successfully made the transition to electronic publishing. Some were too set in their ways to learn a new skill. Change is most devastating to those who can't or won't adapt. What kinds of partnerships will doctors form with expert-system technology? Who will adapt by evolving new caregiving skills that are uniquely human, and who will not survive the blow to their egos?

One lesson that biology and history seem to teach us is to avoid competing with our machines. By making sure that humans and machines occupy distinct ecological niches—that we don't compete for resources or territory—we can survive separately by recognizing and doing what we do best, and they can survive by doing what they do best. Furthermore, by cooperating instead of competing, each can benefit from the expertise and strengths of the other.

But history also shows us how diverse cultures can merge to form a single, new, richer culture. In the longer term, both we and our machines will be able to control the evolution of our own bodies and minds. Eventually, perhaps centuries from now, we will figure out how to liberate our intelligence and our personalities from these bulky organic containers that not only limit our mental capacity but shackle us with troublesome evolutionary baggage. Perhaps we will shed the greed and barbarity that is the legacy of that baggage, while preserving and enhancing our vision, adaptability, and compassion.

17

Beyond Human Dignity

The only inevitability, it seems to me, is that a purpose-seeking animal will find itself as the purpose of nature.

MICHAEL SHERMER

Try to think of an experiment or a test that you could perform that would determine conclusively whether your choices in life are made by an autonomous inner self or are determined solely by a combination of genetic and environmental forces.

Suppose there were such a test, and when you perform it, the answer comes out one way or the other. How would you feel about each possible answer? If you were able to prove the existence of an autonomous self, would you then take more pride in your humanity and your achievements? If you concluded, on the other hand, that your behavior is determined by genetic and environmental forces, would you then feel diminished and less significant?

Now suppose, on the other hand, that no such test can be devised. What would that mean? If there is no way to tell the difference, *is* there any difference?

We are, of course, revisiting the question of free will that we first brought up in Chapter 7. Here, we ask how our human dignity, or sense of self-worth, depends on the belief that our choices are freely made by an autonomous inner being. The way we answer this question will determine how much we respect future intelligent machines and the level of rights and responsibilities we would accord them.

The Ultimate Addiction?

In 1976, Joseph Weizenbaum, the developer of ELIZA, wrote about the moral implications of creating intelligent machines.[1] He saw how deeply attached humans become to their technology (TV, cars, guns, atom bombs) and asserted that machines that enhance and extend human *intelligence* would become the ultimate addiction. This addiction (though he didn't use that word) would ultimately lead to a *loss of human dignity* and many other traits that we hold to be uniquely human—like love, understanding, and interpersonal respect. He felt that we had already (back in 1976!) become slaves to our machines and that we would eventually lose *all* self-respect, if the trend toward the development of intelligent machines continues uncontrolled. His solution was to restrict the expansion of artificial intelligence, forbidding research into certain areas that he thought were the exclusive domains of human beings, including, ironically, psychotherapy.

As we know, Weizenbaum's solution was not implemented, and research in artificial intelligence continues to expand, unhindered by any regulations or other restrictions designed to preserve and protect human dignity. Is there any evidence that the advances in machine intelligence over the last quarter-century have diminished the dignity or self-respect of those who live and work with such machines? Those whose jobs are made obsolete by automation certainly suffer a loss of dignity, if they are unable to adapt. But technology has been doing this for ages, and we accept this as a price of progress.

Although intelligent machines seem to have brought us immense increases in productivity, some people worry about a more widespread loss of humanity, compassion, and self-respect in the information age. One price of increased productivity seems to be that more and more of our daily lives are spent interacting with machines rather than with other people, and many find this dehumanizing. E-mail and Internet chat rooms depersonalize our human interactions, yet they expand enormously the community of individuals with whom we can interact. Virtual communities of people with common interests now span the globe. So, as usual, the new technologies of the information age bring mixed blessings—their abuses inextricably intertwined with their benefits.

As for human dignity, some people seem content to allow machines to do more and more of their thinking for them, whereas others take advantage of the vast opportunities to expand their minds with all the new, readily available knowledge that the information age presents. What one makes of it all—and how one sees his own worth in such an age—still seems to be more an individual choice than an inevitable result.

As for loss of love, understanding, and interpersonal respect, does the information age make it easier to do without these values? Perhaps, but whether we treat each other like humans or machines is still an individual choice. The new technologies at our disposal could just as easily be used to close gaps between cultures and increase understanding as to separate and keep them at a distance. Which path shall we take?

A Science of Human Values?

But isn't there a more important lesson to be learned from all of our misbegotten uses of technology? Excluding from science the *study of human values* condemns us to dealing with twenty-first-century technology, equipped only with rigid moral, religious, and ethical codes that have changed little in most of the world since the Middle Ages. If the human race is to survive its technological adolescence, then maybe it's time to apply our scientific methods to rethink the bases of human ethics and morality. Maybe it's time to develop what B. F. Skinner called a technology of behavior. As a basis for such a technology, we are now, for the first time, equipping ourselves with tools that will allow us to understand and perhaps duplicate some aspects of human thought, while uncovering the origins and structure of our moral and ethical underpinnings. One of these tools is the new science of *evolutionary psychology*, which is providing us with profound insights into the biological basis of human behavior. If the ideas discussed in Chapter 10 turn out to be basically correct—and there is really no nonmagical alternative—then we will have available to us for the first time the means to intelligently modify our own moral programming.

As our knowledge of the social and biological origins of behavior matures, our reliance on myths is slowly being displaced by verifiable facts. Understanding how we work as biological and social beings should better equip us to redesign more rational moral and ethical codes. We may derive comfort from stories about a nonmaterial soul as a way of escaping our mortality. Indeed, the social stability they provide has clear cultural survival value. But stories they are.

What new moral and ethical codes could we design that would be better attuned to life in the twenty-first century? It might be nice, for example, to have social, economic, educational, and governmental institutions that work and even new, less destructive ways to interact with each other. Would moral codes based on reason, rather than our ancestral environment, stand a better chance of accomplishing that? Aren't these the measures by which the value of our science, and our success as a species, will ultimately be decided? Surely all intelligent species (if there be others) reach crucial points in their development, when they figure out certain things about their own nature, that change them forever. Making the connection between sex and procreation gave rise to all kinds of moral machinery to help us cope with that "knowledge of good and evil." Acknowledging the mechanistic nature of man is another of those epiphanies.

If a scientific view of man gave us the tools to change human behavior, what changes shall we make? If we set out to preserve human dignity in the coming age of intelligent machines, what sort of culture could we design? Is there some relationship with our machines that would let us retain our sense of self-respect and uniqueness? Do we need to retain the kind of dignity that comes from a sense of control and superiority, or are these ideas best left behind? What aspects of our humanity do we cherish the most and desire the most to sustain and pass on to future generations? These are all questions about *values*. If we could agree on answers to these questions about our most cherished values, then we might be in a better position to design an environment in which humans and intelligent machines could coexist in a symbiotic way. This is why we need a science of human values.

Nature Plus Nurture

Burrhus F. Skinner (1904–1990) is possibly the most maligned and misunderstood scientist of the twentieth century—often by people who have never read a word of what he has written. In fact, he was, like Charles Darwin, years ahead of his time. Many of the ideas that follow are his. He firmly rejected the traditional view that humans act in accordance with the decisions of a free inner agent. He thought that when a person behaves in an unacceptable way, it is the *environment* that is "responsible" for the objectionable behavior.[2] Therefore, what needs to be changed is the environment, not some mythical internal attitude or other attribute of the individual, such as a "criminal personality" or "sociopathic tendencies."

Skinner's view, which emphasizes environmental influences, seems at first to lie at the opposite end of the nature-versus-nurture spectrum from Darwinian evolutionary psychology, which emphasizes genetic influences on behavior. But Skinner was clearly aware of the genetic effects on organic brain disorders in particular and how our genetic heritage is shaped over the long haul by environmental forces. He simply chose to elaborate on the then poorly understood, shorter-term environmental influences on the brain. When viewed in the light of new insights provided by evolutionary psychology, his works now seem to make even more sense than they did when they first appeared: Human behavior is shaped by both genetic and cultural forces, with the latter being continuously redesigned by man. It therefore seems impossible to disentangle the influences of genes and environment.

Skinner had a (some say utopian) vision of rationally designing cultures in which suitable environmental controls (which he unfortunately called *behavioral engineering*) and incentives would operate on our brains to produce peace, harmony, and productive growth. Before we would ever accept such controls, however, he thought that we would have to abandon our cherished "values" of freedom and dignity altogether.

Those who see human history as a continuous struggle for freedom and dignity find Skinner's *behaviorism,* as it is called, hard to swallow. Most people find the idea of intentional social control, or behavioral engineering, ethically and morally repugnant. Their opinions are no doubt biased by the legacy of such aversive methods as those used by Nazi Germany, those depicted in the futuristic novels *1984* and *Brave New World,* and those shown in Stanley Kubrick's film, *A Clockwork Orange.* They see control as the opposite of freedom—always bad because, in its crudest forms, it produces undesirable consequences, such as slavery and totalitarianism. But must this always be so?

It is sometimes said that you can't design a culture scientifically because people simply won't accept being manipulated. "I wouldn't like it!" is the simple rejoinder. But consider all the "benign" social controls that we already willingly submit to—the ones imposed by governments, religions, advertising, political propaganda, military training, the media, and popular culture. Do these controls seem less insidious because they are more subtle and not quite so deliberate and because we are free to ignore them? Think again! What parts of our minds and our precious freedoms have we relinquished to these controls?

Deliberate environmental manipulation is widely practiced in all cultures—by parents, teachers, governments, employers, therapists, and religions, to name a few. A trivial example of environmental manipulation would be closing a window when we feel cold. Another is the spurious "needs" put into consumers' minds by advertising. A complex one would be the military training that conditions people to hate an enemy and to sacrifice their lives for their country. Why do some kinds of control seem morally and ethically acceptable? Perhaps they seem acceptable because they deliberately create the illusion that we have some *choice* about accepting the controls.

The Limits of Freedom

Because he rejected the idea of autonomous man, Skinner regarded freedom, not as a *cause* of behavior but as merely an *imagined precursor* to it. He wrote that our culture's extensive "literature of freedom and dignity" has contaminated our thinking about the origins and causes of "good and bad" behavior and about useful ways to change it. His scientific view of man offers new possibilities for eliminating many undesirable behaviors. It should be possible, he says, to design a world in which behavior likely to be punished would seldom or never occur. But people reject these views because they fear the loss of freedoms they have been conditioned to regard as precious.[3]

Happiness has long been associated with the freedom to do what we please. *But where are the boundaries beyond which doing what one pleases may threaten the survival of a culture?* On which side of that boundary lies the freedom to use up the earth's limited resources? The freedom to reproduce as often as we please? The freedom to carry and use lethal weapons? The freedom to take advantage of the less fortunate? The freedom to turn the air, land, and sea into garbage dumps? The freedom to change the earth's climate in unpredictable ways? And now we are forced to ask the same question about the freedom to pursue any line of scientific inquiry we please.

The problem has always been knowing where our freedoms end and our responsibilities begin. Will the Great Democratic Experiment ultimately fail because it forgets this elementary truth that its designers knew so well? Individual and corporate greed cannot long flourish at the expense of society at large, future generations, and planetary integrity.

The End of Human Dignity?

We recognize a person's dignity or worth when we applaud an accomplished artist or musician, a skilled athlete, an astute businessperson, a brilliant doctor, scientist, or engineer—or even a child who eats her vegetables. Generally, we praise and commend people when they behave in a way that we approve of. Praise is pleasurable because we have learned that it is often followed by other rewards. On the other hand, when we are reprimanded or criticized, we tend to lose self-respect and fear the punishments that might follow.

The way that we administer approval and disapproval has evolved over the ages—in Western cultures at least—in such a way that we assign credit and blame to an autonomous individual who alone is responsible for the behavior. (Eastern cultures harbor fewer illusions about control.) But the more we learn about the effects of genetic and environmental forces on behavior, the fewer parts of that behavior we can attribute to an autonomous internal agent. A scientific explanation of behavior leaves no trace of autonomous man and nothing for which they can take credit. But if we do away with the idea of the autonomous individual, what happens to human dignity and self-worth? If we suggest that commendable behavior results from our genetic programming and cultural environment, and not an autonomous individual making free choices, then many people feel diminished and insignificant. The question is whether building the concepts of dignity and self-worth on the idea of an autonomous individual is beneficial or destructive. In other words, do we have to believe we are autonomous to feel good (or *feel good about ourselves*, in psychological lingo)? If not, what is the alternative?

How about this? Human beings are an amazing and wonderful (and to some, a seemingly impossible) construct of nature. Our brains and nervous systems are marvelous machines that transform a mass of sensory inputs into the incredible variety of human behaviors. That a machine of such complexity and elegance could possibly exist should be an infinite source of wonder and inspiration—and a much more satisfying source of pride and dignity than some magical, unexplainable inner being.

To be sure, undoing a lifetime of conditioning that ties our dignity to our autonomy is a tall order. Does it mean we should withhold praise for desirable behavior? Certainly not! Praise and other rewards (reinforcement) play an important role in assuring that desirable behavior is repeated. It seems universal (that is, genetic) that praise and compliments make us feel good and cement social relationships as well. If so, so much the better. We need not respond to compliments like *Star Trek*'s Mr. Spock, who typically dismisses them as superfluous.

My Genes Made Me Do It!

If we recognize free will as illusory, then where does that leave us on the weighty issue of accountability, morality, and ethical responsibility? Does it mean that our future is determined and therefore beyond our control? Are blame and credit then meaningless concepts? Does it make sense to overhaul all our moral and ethical codes and replace them with "no-fault" versions? Or should we just continue *acting as though* we do have free will and are morally accountable for our actions? Or maybe we should take a pragmatic view and ask how well we think our existing codes work—or rather which parts work, and which parts don't—and patch up the parts that don't. Perhaps our minds are designed to rethink our moral and ethical codes at just the same time that we begin to uncover the illusory nature of free will.

Understanding that our choices are not free and arbitrary does not automatically free us from accountability. If what we "choose" to do is really the sum of the subconscious influences of genetic programs, experiences, knowledge, and conditioning, then these are surely the ingredients of our "responsibility." If we are not careful, however, this line of reasoning can lead to the "My genes (or my upbringing or my diet) made me do it!" defenses against any criminal charge. Although these defenses may be technically correct, wouldn't they, if allowed, hopelessly subvert a legal system that insists on assigning guilt or innocence to the individual?

A medical analogy may clarify the point. Modern biochemistry is discovering the genetic basis for many diseases and birth defects, but when a faulty gene is discovered, we do not pronounce it guilty and incarcerate or kill it. We try to find out how the genetic fault occurred and prevent it from happening in the future. Sometimes, understanding the genetic basis for a disease may lead to useful treatments, including genetic manipulation. Similar arguments can be made for physical disorders of environmental origin. Understanding the toxic effects of certain chemicals, for example, typically leads to regulations governing their safe use.

We do not yet know how to "rewire" a brain that has been culturally programmed for, say, criminal behavior, but understanding the environment that produced it can lead to useful remedies. Offenders can acquire new knowledge or experiences that will help them make better choices. What's more important, such an understanding would focus attention on underlying environmental (cultural) causes and on changes designed to minimize the incentives for criminal behavior in future generations. This science is still in its infancy, and our primitive desires for punishment and revenge will always stand in its way. But this approach certainly offers more long-term hope than the present penal system.

Who's in Control?

Each culture in which we live is packed with intentional behavioral controls of all kinds, about which we have little choice. Public education and military training are examples. But we are so used to them that we barely notice them. Consequently, we should expect a lot of resistance to any attempts to institute different controls and incentives. Even if they promise to create a cultural environment that offers much greater rewards, the refrain *I wouldn't like it* would probably crop up time and again. You can hear the naysayers now: *Oh yeah? What kind of controls and incentives? Will I still have control over my life?* But the problem is not to create new controls that would be liked by those who live under today's illusions of freedom and self-determination. Perhaps the best we can do is reconstruct the values we pass along to future generations.

As the emphasis gradually shifts toward restructuring our values and environment, new fears and concerns will arise: Who is to construct the controlling environment, and to what end? Of course, no one knows the answers to these sticky questions yet—the science of human behavior is still in its infancy. Eventually, we may learn enough about cultural design to produce certain desired results. Then the question will be, What results do we want? One goal might be to give people rules to follow, so that they might avoid nature's punishments, live together in peace and in balance with nature, promote cultural growth, and so forth. But when a *science of human values* matures a little more, we may find better goals and rules for getting there. Such rules would not destroy freedom, responsibility, or any other mystical quality. They would simply make the world safer and more rewarding for everyone. As Dean Wooldridge said: "Society profits when its members behave more intelligently. And men who know they are machines should be able to bring a higher degree of objectivity to bear on their problems than machines that think they are Men."[4]

Skinner and Wooldridge could have phrased their ideas in other more palatable ways that might not have provoked such vehement objection. In 1954, Dorothy Law Nolte, a contemporary of them both, wrote her oft-quoted inspirational poem, "Children Learn What They Live," which expresses in plain language the controls of environment over behavior. The poem includes these lines:

> *If children live with criticism, they learn to condemn.*
> *If children live with hostility, they learn to fight.*
> *If children live with honesty, they learn truthfulness.*
> *If children live with fairness, they learn justice.*[5]

We can easily make up other lines that follow the same reasoning: If children live in a culture that glorifies violence, they learn to solve problems by violent means. And if children live in an environment that values money and material possessions above all, they learn greed as a way of life.

If Skinner had expressed his behavioral-engineering ideas in this warm and fuzzy form, then perhaps they would have been more widely appreciated and accepted.

18

Extinction or Immortality?

To the person on the street, the most troubling question about artificial intelligence is *If we build machines that are smarter than we are, then what will become of human beings?* Science fiction has come up with many possible, even plausible, scenarios, ranging from extinction to immortality, with various states in between. A popular sci-fi theme has predatory war machines in charge of a post-apocalyptic earth, hunting down and exterminating the last of the rebellious humans (e.g., the movie *Terminator*). Another has humans living in harmony with many kinds of intelligent but subservient robots (e.g., *Star Wars*). Still others portray humans that have escaped their carbon-based vessels and exist as disembodied or communal minds in more durable containers (e.g., *Star Trek*). (Notice that the status quo is not considered an interesting option.) Are the next steps in human and machine evolution more likely to take us down the path to extinction or to immortality? Or is there really any difference?

The Search for Immortality

In 1986, psychologist Robert Ornstein observed that, for all practical purposes, we are the same people who lived in small groups, roamed

the savannas of east Africa, and faced daily threats from wild predators.[1] Our bodies and brains evolved to suit the world of 20,000
years ago, and neither has changed significantly since then. Those
same bodies and brains are now trying to meet the challenges of a
world that changes dramatically in a lifetime. In a process that Ornstein called *conscious evolution*, humans both create and adapt to
these changes, using technology to greatly amplify and extend the
capabilities of our muscles, our senses, and our minds. So our tools
are getting smarter, but we are not!

We even use technology to extend our own life spans, by controlling disease and by replacing aging organs with borrowed human
ones and even mechanical ones. Still, only a very few people live
more than 100 years, and no one we know of has lived 150 years,
even with mechanical organs. Will we eventually be able to extend
life indefinitely by replacing more and more worn-out body parts
with synthetic equivalents, until we get something like the six-mil-
lion-dollar man in the 1970s television series by the same name—a
kind of *Homo cyberneticus*?

But wait! Instead of tedious, piecewise replacement, why not just
replace a whole worn-out body with a stronger, better-designed,
and longer-lasting one that acts as a container for the brain? Brand
new and improved limbs and sensors would be connected to the
same old personality and "self" in our brains! We wouldn't even
need hearts, livers, kidneys, or gall bladders—they would be replaced with more reliable electromechanical systems. The messy
part would be preventing the natural deterioration of the brain. Do
we really need that? A common sci-fi theme supposes that our
human intellects—the entire electrical contents of our brains—
could be "downloaded," bit by bit, into a more durable machine, allowing our minds and consciousness to far outlive our biological
bodies.

Get rid of the idea of *biological bodies* altogether! Robotics scientist Hans Moravec believes that one day we will be able to replicate

an entire human being—or for that matter, a whole community of human beings—in a computer simulation.[2] We can't imagine today how such downloads would work. We know, for example, that the brain is an *electrochemical* system, not just an electrical one. Perhaps some kind of high-resolution neural tomography—a vastly more sophisticated kind of brain scan than present-day computerized axial tomography (CAT) and positron emission tomography (PET) scans—could map the electrochemical activity of the brain on a cellular level and from millisecond to millisecond. But even if we do figure all that out, would the resulting entities in any sense be human? Suppose *you* were downloaded into a computer. Just think for a minute about how you might spend your time. So for now, the question about this kind of immortality is not only *Is it possible?* but also *Would we even want to do it?* For my money, preoccupation with this sort of immortality for the benefit of a few wealthy humans would be a waste of time and scientific resources. There are just too many more fruitful ways to improve the human condition.

Another road to immortality takes more direct and conscious control of our *biological* evolution using genetic engineering. Once we figure out what the structure of our DNA means and how our genetic program works, we should be able to develop new cancer drugs, find ways to grow replacement organs, control aging, and discover new ways to prevent disease and birth defects. An advantage of genetic engineering is that we could simulate all these processes on a computer before trying them out on living people, thereby reducing the chances of huge genetic blunders!

We can therefore embark on two seemingly distinct paths to immortality: Following one path, we would supplement and eventually replace our organic parts—perhaps including our brains—with mechanical ones. On the other path, we would control and even design the configuration of our own organic bodies and minds from scratch. It seems likely that these biological and mechanical paths will eventually merge, as nanotechnology matures, permitting us to manipulate and supplant our organic parts on cellular and molecular levels. Then we should be able to engineer any kind of organic modifications we wish—even design entirely new human beings!

But once we get into this business of playing God, why stop there? If we can design human beings, why make any more of the greedy, violent, barbaric, self-absorbed kind? Why not nicer, testosterone-free, superhuman beings? Or entirely different kinds of intelligent life altogether? There is certainly no shortage of examples of desirable human attributes and powers in our history that we could nurture and amplify, just as there are many that we could better do without. This is where the difference between extinction and immortality blurs.

Extinction or Evolution?

We have already discussed in Chapter 14 the chances that intelligent machines might outsmart and eventually replace human beings as the dominant life-form on the earth, and we will discuss in Chapter 19 our unwitting complicity in any such takeover. Whether we regard such replacement as the extinction of the human race as we know it or as just the next stage in our evolution may be simply a matter of personal taste. After all, extinction is an essential tool of the evolution and improvement of every species—pruning the evolutionary tree, if you will. And some would regard the end of humanity in its present form as a blessing for the planet, since we've made such a mess of the place.[3]

People have always tried to improve themselves, and now they are on the threshold of a major renovation. The possible nature of *post-biological man*, as transformed by technology, is debated in an interesting discipline called *transhumanism,* which you can follow on the Web.

Does all this sound too much like science fiction? The difference is that these events have a significant chance of actually taking place, if we don't destroy ourselves first. The vision of science-fiction writers often gives us a peek into the future. Some are pure fantasy, but some science fiction has a way of becoming science fact. It's not always easy to distinguish between the two in advance, except by asking whether they violate any laws of physics. The visions expressed in Jules Verne's *From the Earth to the Moon* were realized in 1969. *Flash Gordon*'s ray guns have become today's laser weapons. The communicators used in *Star Trek* have turned into today's cell phones. The intelligent machines of the future may be more or less like those portrayed in science fiction, or something totally unexpected, but what is not in question is that we have already set out down that road. Any sci-fi fan knows that the big problem, pointed out nearly two centuries ago in *Frankenstein*, is losing control of our creations.

Where Is Everybody?

We cannot discuss the possible extinction of humanity and the emergence of intelligent machines without uncovering some really big questions, such as *Is extinction the inevitable course of events wherever intelligence and technology evolve?* Although this question may seem unduly esoteric, it does have some practically important consequences for evolutionary biology, space exploration, and the search for extraterrestrial intelligence, not to mention science and technology policy in general.

By most estimates, the universe is teeming with intelligent life. Conservative guesses place thousands of technologically advanced civilizations within radio earshot of Earth. So why don't we see any signs of other intelligent civilizations? Shouldn't we have already heard from them, or at least overheard some of their cell-phone chatter? And why haven't we been visited by curious extraterrestrials? The question *Where is everybody?* has come to be known as the Fermi Paradox, after atomic scientist Enrico Fermi, who posed it in a seminal 1950 article.[4] Years of active searching since then, with the world's most powerful radio telescopes, have failed to detect any verifiable intelligent radio signals from outer space.

Lots of reasons for this negative result have been suggested. Maybe we're the first intelligent civilization in our galactic neighborhood. Someone has to be. This is possible, but highly unlikely. Maybe intelligent and technological civilizations aren't as common an evolutionary outcome as we think. After all, the dinosaurs lived and evolved for 140 million years without developing technology. Maybe communication by radio broadcasts is only a brief and primitive interlude in the evolution of technology. Individuals in an even slightly more advanced civilization would likely develop broadband mind-to-mind connections that would be hard to listen in on. Maybe mass-extinction events, like asteroid impacts, are much more common than we think. Or maybe extraterrestrials have some kind of prime directive about interfering with primitive civilizations like ours. And so on.

The gloomiest explanation of the Fermi Paradox relates in an interesting way to our investigation of the moral and ethical implications of thinking machines. The reason we have not encountered signs of any other intelligent life-forms, this explanation goes, is that technological civilizations have a very short life expectancy, because they promptly destroy themselves during their technological adolescence. We can already see plenty of ways that this could happen: nuclear or biological warfare, runaway climate change, overpopulation, blunders in genetic engineering, an explosion of nanobots. More subtle forms of extinction would accompany the systematic destruction of the biodiversity of the planet, on which human life depends. It is difficult to imagine how any technological civilization that allows an exponential proliferation of more and easier means of self-destruction can possibly keep them under control indefinitely. Won't someone slip up sooner or later, like the pilot testing a new rocket fuel in Kurt Vonnegut's *Slaughterhouse-Five*?

Opponents of this grim view say yes, maybe so, but even if most technological civilizations do quickly burn out, that will still leave a huge number of "technologically wise" survivors to proliferate and colonize. If so, are there any ways to assure that our species is among those lucky few that will acquire the necessary wisdom before doomsday strikes?

Generally, humanity is not very good at reacting as a whole to abstract threats in the indefinite future. The danger must be perceived as real, immediate, and personal, and there must be general agreement about what to do. Large-scale warfare is probably the closest we come to rallying behind a single cause. Possibly, a very close brush with extinction in the form of a big, but not apocalyptic disaster, would catalyze global action. But what kind of action?

What we seem to need is enough foresight or wisdom to help us *steer safely* through the rocks and shoals of technological development. Maybe the technologically wise civilizations survive by using their advanced technology to *model alternative futures*. Such models would allow them to see the consequences of different technological developments and to test new codes of behavior. Accurate simulations would simply enhance the foresight and the planning and organizing abilities that have only recently evolved in our brains' frontal lobes. They also seem essential to insure that technology and moral codes keep pace with each other, so that technology doesn't outrun a civilization's ability to deal with it.

Peaceful Coexistence?

Some see a rosy future of peaceful coexistence with machines that, in spite of their superior intelligence, would still treat us with benign respect. After all, if they are smarter than us, then they will certainly have a more enlightened attitude toward lower life-forms than we do, right? Might they even regard us as helpful, if somewhat retarded, partners?

Possibly, except for one crucial fact: a lot of people will deeply resent being displaced as the dominant life-form on earth and will certainly fight to the death to defend that status! Ranks of neo-Luddites will surely expand, rising up to sabotage the proliferation of intelligent machines. Technophobes, who today believe they have the option of ignoring what is going on and simply dropping out of the technological revolution, will sooner or later feel compelled to abandon their passive resistance and take up arms against the machines. If open warfare between people and the machines ever breaks out, there is little doubt who will start it.

How will the machines react to this threat? Probably much the way the computer Colossus did in *Colossus: The Forbin Project* (Chapter 14). By that time, the power that we have given machines over our lives will be so far-reaching and pervasive, and we will be so dependent on them for our everyday subsistence, that they will literally hold over us the power of life and death. They will not have to seize power; they will already have it! What aspects of our lives have we not already handed over to computer control? In the future, our supplies of food, water, heat, and electricity; our communication networks; our financial and transportation infrastructure—all will have been willingly turned over to the control of machines. Today, we may be able to limp along when our computers go down, but soon, no one will remember how to work these systems without the computers. Elaborate safeguards against human meddling—initially put there by us to protect critical systems from sabotage—will, ironically, prevent us from taking back manual control, which no one will know how to do anyway!

For all these reasons, it would appear to be in the machines' best interests to do away with those pesky humans, and furthermore, they would probably have no problem doing so. Before we face this fate, however, we still have choices that might allow us to avoid this outcome. So what are those choices?

19

The Enemy Within

Artificial stupidity may be defined as the attempt by computer scientists to create computer programs capable of causing problems of a type normally associated with human thought.

WALLACE MARSHALL

Although our fears that intelligent machines will someday take over might be justified, I don't believe they can do so without our complicity. *The greatest threat to our dignity and our humanity will not come from machines that act like people, but from people who act like machines.* The more we allow ourselves to be told what to think and to be treated like automata by other people, by governments, corporations—and yes, by computers—the more vulnerable we become to domination by intelligent machines.

The visions of *1984*, *Brave New World*, and other dystopian prophecies have given us chilling previews of a world in which people are transformed into zombies because they have surrendered their minds to some higher power. Orwell's and Huxley's warnings merely extrapolated trends that they saw in the totalitarian societies of their times.

Such trends are not new. Whenever one culture sets out to control or enslave another, the first steps are always to control information and to enforce a machine-like submission to authority. By forcing people to give up reason, and with it their humanity, conquerors can justify treating other human beings like cogs in the machinery, or worse. When we lose our courage to ask questions, when we let others do our thinking for us, when we allow our ideas to be muzzled by political correctness, we relinquish our humanity, not just to tyrants and demagogues, but to bureaucrats, politicians, the media, advertising, military paranoia, and religious dogma.

Surrendering Our Minds

We begin to surrender our minds when we meekly follow those in authority or accept ideas without question, simply because they sound appealing or because that's the path of least resistance. Hitler secured a vast following in depression-ridden Germany by invoking its citizens' sense of national pride and purpose. But in time, mechanical adherence to Nazi propaganda inexorably turned ordinary citizens into collaborators in the subhuman atrocities of the Holocaust. "Just following orders" became an acceptable excuse after the war, because it shifted guilt from the individual to a ruthless and autocratic government. Apparently, moral responsibility is suspended under the influence of authorities sufficiently powerful to capture and manipulate our minds. Hitler himself is supposed to have said, "How fortunate for governments that the people they govern do not think."

With their racist propaganda, the Nazis mastered the kind of mind control that would later be called *groupthink*. This term was coined in the 1970s to describe how we abandon our judgment and our perceptions of right and wrong to a common group or herd mentality that reinforces normally unacceptable behavior. Classic examples are a lynch mob, a street gang, a college fraternity, a stadium full of sports fans, a political convention, and a military unit under the stress of battle. The group need not be small or consist of personal acquaintances to seriously distort values and morals. Corporations and nations just as effectively condition us to conform to their policies. Religious cults induce bizarre rituals, such as the Jonestown mass suicide.

Symptoms of groupthink are feelings of invulnerability, being oblivious to consequences, not being accountable for actions, hostility toward nonbelievers, and suppression of individual feelings and sensibilities. Criticism is suppressed as disloyal. Groupthink works because we all feel safer and more willing to take risks when the risk is shared—when "everyone else is doing it."

Do you recognize any of these symptoms of groupthink in your daily lives at home or in the workplace? Which pressures to conform do you submit to without thinking? Do peers and the media dictate how you dress, what you eat, what car you drive, what music you listen to? Are you a team player, or do you question assumptions? Do you go along with the majority, fearing that disagreement might be seen as disloyal? Do you compromise your beliefs for the sake of consensus and harmony? Are your moral and ethical principles for sale for a few bucks?

Waiting for the Messiah

When the world seems to be going mad, we traditionally look for some kind of messiah—a savior who will bring the wisdom to rescue us from all the social disasters that we bring on ourselves. When we look back through history at individuals who laid claim to that title, we find people who mainly tried to tell us what to think, by virtue of some special insights they possessed. Perhaps in those times, special insights were what civilization needed. If the ground were not fertile and receptive, their preaching would certainly not have taken root.

Today, we face the opposite problem: There's no shortage of messiahs. So many people—governments, corporations, educational institutions, the media, religious institutions—are trying to tell us what to think, so that the challenge now is figuring out how to think for ourselves. To this already long list of authorities to which we so readily yield our humanity, we must now add intelligent machines. We already allow bureaucrats to blame computers for their own ignorance, inflexibility, or incompetence. How many times a day do you deal with people who respond to your requests or inquiries by saying something like *I don't know—I just work here,* or *That's just our policy,* or *I'm just following orders,* or *That's what the computer says.* Will this only get worse as computers take over more and more of our affairs and relieve us of more and more of our physical and mental challenges? Are there ways to coexist with intelligent machines in a way that allows us to continue to grow and learn? Or will we allow control to gradually slip away?

These questions are as old as technology itself, but they take on new meaning as messiahs of our own design not only relieve us of physical drudgery, but also offer to do our thinking for us. If intelligent machines think us out of our jobs, and if unemployment becomes as widespread as some envision, will the bulk of the work left for humans consist of servicing the machines? Or will we face no greater challenges than deciding how to spend our leisure time, where to shop for the latest stuff we don't need, or where to find still more stimulation for our bored minds and bodies?

The New Slavery

Before machines did most of our work, it was done by animals and slaves. Until the Industrial Revolution, slavery was socially acceptable because there was no alternative source of labor. Not until viable technological alternatives appeared was the morality of slavery seriously questioned. Today, slavery in its traditional form is common only in less industrialized cultures. But has a new, much more subtle form of bondage that enslaves people's minds taken its place?

Back in 1956, long before the term *yuppie* was coined, Sloan Wilson's *Man in the Gray Flannel Suit* called attention to the subservience and mind-numbing conformity of the corporate world. In the film version of the book, Gregory Peck portrayed a young Madison Avenue ad executive whose life is torn between his disintegrating marriage and the pressures of escalating company responsibilities. That such conflicts even existed was a shock to the *Ozzie and Harriet* generation of the fifties.

Today, we more or less accept that large corporations often treat their employees like indentured servants and their customers like prey; that government bureaucracies are likely to be impersonal, unresponsive, and rigid; that autocratic managers often dictate policy by remote control and create workplaces that foster adversarial and competitive relationships between employees. Battlefield terms are often applied to the workplace: *stick to your guns, employees in the trenches, stay on target, we need more ammunition, road warriors,* and so forth.

An even more disturbing trend in industry is to reward employees for their abilities to get along, not with people, but with machines. In fact, more and more of our lives are spent interacting with machines, rather than with other humans. Assembly-line workers have known these pressures since the days of Henry Ford. Today, résumés are scanned and parsed by machine. More and more workers have their time managed and their performance measured by machines. Devices record the time employees spend at their work stations, the number of work units processed, or the number of keystrokes executed. Their e-mail and Internet transactions are monitored by machines. Leadership is largely replaced by project-management computer programs, and people spend their time generating reports that no human ever reads.

Look around the next time you are at an airport. Notice how many people are tethered to their worlds by cell phones, personal digital assistants (PDAs), and notebook computers. Who's in control, the machines or the people?

Who Thinks for Us?

For slavery to work, the masters have to *control what the slaves think* by strictly managing their access to information. If you completely control a people's access to information, you can make them do virtually anything. You can change their worldview, their self-view, even their sense of right and wrong. Patricia Hearst, held captive by the Symbionese Liberation Army for more than a year and a half, apparently underwent a transformation from socialite to SLA soldier and bank robber, and then back again. In the antebellum South, teaching slaves to read was a crime. During the Cold War, the United States and the Soviet Union conditioned their citizens to fear each other as mortal enemies and to spend trillions of dollars to assure each other's destruction.

In many countries today, access to outside media is tightly controlled, lest their citizens' minds be exposed to unorthodox ideas. In "free" countries, information is managed in more subtle ways. Corporations and government agencies filter and sanitize information for their employees and for the public, to lull people into a sense that everything is just fine. Mass media make a living managing information for us, sorting it into bite-size chunks, so that we know less and less about more and more. Most of us condemn the brainwashing suffered by many prisoners of war, but we accept as a normal by-product of capitalism the omnipresence of modern advertising that pollutes our environment and numbs our minds.

Now the World Wide Web is the vehicle of choice for disseminating all kinds of information, as well as disinformation. In one edition of *Sixty Minutes*, Lesley Stahl was interviewing a guy who runs a Web site about conspiracy theories, a subject about which humans are notoriously receptive. He fabricates and puts on his site all kinds of wild stories, interweaving enough truth with his fantasies to make them seem credible. An appalled Stahl confronts this guy, asking if he feels any responsibility to tell the truth. If people like him just put anything they want on the Web, without making any distinction between truth and opinion and fantasy, then how, she asks, is the public ever going to sort it all out and find out what the truth really is? Shouldn't there be some controls on the Web to prevent people like him from irresponsible reporting?

What Stahl failed to realize was that having someone decide for us what is responsible and irresponsible reporting is just as dangerous as the misinformation itself. (Did she have in mind that only professional journalists should be certified as the custodians of the truth?)

Many people consider the Web dangerous because it is the first completely free and uncensored global information channel. Every point of view on earth is represented, including some that we may find offensive and even terrifying. The price we pay for the almost infinite amount of information and misinformation on the Web is that *we have to decide for ourselves* what to believe and what not to believe! Many of us are so used to having our information managed for us that we have no tools for sorting the truth from the rubbish.

The Lost Art of Critical Thinking

People have different ways of deciding what information to believe. One way is to believe what they are told by "usually reliable authorities"—teachers, governments, journalists. For some, the authority is now the Web—if you see it on a computer screen, then it must be true.

The second way is when people believe what they want to believe—that is, whatever reinforces their worldview and makes them feel good—regardless of the evidence. This is usually called *faith*. Arguments based on reason are unpersuasive when people decide what to believe based on emotions.

A third way is to ask certain questions designed to sort fact from fiction. This way is called *critical thinking*. For example, if someone claims to have been abducted by extraterrestrial beings, you might accept the person's claims at face value if you have already decided that extraterrestrials regularly visit Earth. On the other hand, you might ask for proof of the abduction, such as an alien artifact, or you might ask yourself what the person stands to gain by making such a claim.

Just when we seem to need it the most, clear, critical, logical thinking seems to be going out of fashion. As a result, huge numbers of people believe in astrology, extrasensory perception, ghosts, the devil, angels, telepathy, psychic readings, reincarnation, communication with the dead, the healing power of crystals, and extraterrestrial visitations. Critical thinking is seldom taught in schools—the kids might upset classroom order by challenging their teachers! Popular culture portrays as nerds and negative thinkers those who think critically. A TV commercial for the Bank of America featuring Charles Lindbergh unfairly portrays skeptics as naysayers who consistently inhibit innovation by saying *It can't be done* and *It'll never fly*.

In truth, skeptics perform a valuable service by questioning revealed wisdom, by examining unfounded claims, by challenging pseudoscience, and by urging people to think more critically about the information they are fed. Might Madison Avenue have some motives to squash this ability by defining it as uncool?

Of course, people are free to believe whatever they wish. So, what's wrong with holding on to comforting beliefs, such as an afterlife, that may help us cope with stressful events, like the loss of a loved one? Myths get started because we take comfort in them, and because they serve a useful purpose in a species that likes explanations for things. Such beliefs seem harmless on the surface, but large doses of them condition our minds to ignore reason, to "trust our feelings," and to accept revealed wisdom on faith. What does it matter, unless you're an engineer or a physicist, that there is a physical explanation for the colors of a rainbow or the glow of a sunset or a broken heart? And myths are so hard to let go of that a person might fairly ask what harm is done by holding onto them.

The answer, I believe, can be seen in the world around us. How much of what you see makes sense? Social, economic, educational, penal, governmental institutions that don't work. Out-of-control population growth that endangers the planet's health. Cultures dominated by greed, violence, and self-absorption. People killing each other in the name of God. What harm, indeed!

Critical thinking, on the other hand, provides ways to test and validate the information we are exposed to. If you want to know whether the earth is flat, if there can be perpetual motion machines, if witch doctors can cure cancer, or if global warming is real, there are fairly straightforward ways to find out. Critical thinking also frees us to question our own values and beliefs, when our lives don't seem to be working the way we want.

In the modern world, it's impossible to subject every claim we encounter to critical analysis—it would simply consume our lives. With this in mind, you can take some simple shortcuts that will get you through many everyday situations: First, it is unwise to trust claims made by someone who has something to gain by their acceptance. (Never send the Defense Department to assess an enemy threat.) Second, question any truths based on an appeal to a higher authority ("It says so in the Bible"). Third, question any claims that appeal to tradition. ("Everyone else thinks so" or "That's the way we do it here"—remember groupthink?) For more extensive lessons in critical thinking, I recommend two excellent books, *The Demon-Haunted World* and *How to Think About Weird Things*.[1]

Finally, you can subject claims to critical analysis only if the information is available. Secrecy, of course, is widely used to keep people from doing just that. This is another way that governments and corporations think for you and perpetuate activities—often ones you pay for—that would not withstand public scrutiny. *Glasnost*, or openness, gives us a say about whether each new technology that comes along is transforming our way of life for better or worse.

If human dignity means anything, then it surely refers to an ability to chart our own destiny—to have something to say about the world in which we live, work, and raise our children. The enemy within is our willingness to let others think for us—to surrender control over our lives. Just as we have lost our ability to do long division to pocket calculators, we are losing our ability to make intelligent decisions about significant aspects of our own lives. A popular bumper sticker says: IF YOU'RE NOT OUTRAGED, YOU'RE NOT PAYING ATTENTION! The challenge now is to find ways to use intelligent machines to make our lives and our work more productive and satisfying, instead of allowing them to turn us into their witless servants.

20

Electronic Democracy

Technology helps democracy by eroding secrecy; but technology hurts democracy by eroding reflection and time. Yes, the people usually know what's best. It just takes us a little longer than the push of a button to figure it out.

JONATHAN ALTER

Whatever happened to all those promises from the 1970s, that technology was about to remove all the obstacles of distance and time and make democracy more efficient and truly participatory? In the twenty-first century, we were all supposed to be voting from home, taking part in instant electronic polls, providing instant feedback to policymakers, and participating in national "town meetings."[1] Ross Perot even campaigned for president in 1992 on a vision of *teledemocracy*.

One reason these predictions have not come to pass is that they underestimated the average voter's political apathy and conservatism. Among even the most liberal voters is a widespread fear that teledemocracy would exclude large sectors of society, mainly the poor and the technophobic, from participation. Like the paperless office, teledemocracy seems as distant a promise as ever. Even so, information technology has yielded expanding dividends for democracy in ways that could not have been foreseen before the Internet.

We often hear that the information revolution is about a global redistribution of power—a kind of decentralization that will let "The

People" take control back from governments and corporations. But if more individuals get involved in decision making, would the result be consensus or chaos? Could intelligent machines bring order to the process? Could they help manage the flood of information that citizens need to process to keep abreast of issues? And by assuming the role of supernegotiators, could they reconcile diverse interests, moderate complex discussions, and help reach consensus?

Do the People Really Know What's Best?

George Bernard Shaw said, "Democracy is a form of government that substitutes election by the incompetent many for appointment by the corrupt few." It's easy to be cynical about all the aspects of democracy that don't seem to work as advertised, but as Winston Churchill said, "Democracy is the worst form of government—except for all the others." Maybe there are ways to patch it up and make it work better.

Information technology is capable, in principle, of permitting full and direct citizen participation in government at all levels, from local to global. The question is, *Is that what we want?* Most of us would like more say about certain issues of direct concern to us. We want to give policymakers useful feedback about where we are going, where we *want* to go, and why. We want to freely discuss these issues with others who have like concerns.

The trouble is that few people have the time or inclination to participate on a sustained basis. There's just too much information to absorb and analyze while leading a normal professional and family life. In addition, many issues are just too complex for the nonspecialist to grasp. Even the founding fathers realized this when they set up a representative form of government. Having more say in government and leading a normal life seem like conflicting goals. Can information technology help create a more informed populace, without disrupting their lives and overwhelming them with information?

Suppose there were a computer system that had access to all the public information about pending legislation, public policy issues, and the positions of every elected official and candidate for office. Using intelligent search bots, it would have access to the ongoing public record of everything that goes on at all levels of government, suitably indexed and cross-referenced.

Suppose also that this system could conduct a strictly confidential interview with you. The interview would be thorough enough that the system could construct a comprehensive profile that represents all your important values, interests, principles, philosophy, moral codes, and political leanings. You could update your profile and also provide your positions on specific issues whenever you wished. You would naturally need credible assurances that your profile could not be tampered with or used in unauthorized ways.

Equipped with all this information, our intelligent computer system would seek out and prioritize for you all the issues under current debate that match your interest profile. It would be smart enough to strip out all the officialese and legalistic gobbledygook and to summarize key points clearly and concisely, with hypertext links to more detail, if you wanted it. It could also set up contacts with key policymakers and let you join debates on issues of interest to you. If you told the machine how much time you had to devote to an issue, it would set up discussions with other citizens who have different perspectives on the same issue. It could draft letters for your approval that communicate your views most effectively to policymakers. If you chose a minimal level of effort, you could even authorize the machine to *think and vote for you* on issues important to you, in a way that is consistent with your profile. Of course, some people would see this option as dangerous Big Brotherism that further relinquishes control to the machines. Still, it might be better than no involvement at all.

Another role for intelligent machines in teledemocracy would be a kind of super-negotiator. The more citizen participation, the more the decision-making process will get clogged up with diverse values, interests, and views. Debates and discussions would more often become chaotic and disintegrate into rigid, warring factions—pro-life versus pro-choice, for example. An electronic democracy would have to find efficient ways to unclog debate, while giving people the feeling that their interests and views have been taken into account. Although many of the principles of effective negotiating are well established, emotions often contaminate the process, and very few humans seem to negotiate well.[2] An intelligent machine could be taught all the principles and tactics of successful negotiation. It could moderate discussions by finding common ground and by creating options that almost everyone can agree to. (Taking the process one step further, could it even replace our elected representatives?)

To test our negotiating machine, it might be instructive to try to reach a popular consensus about wise directions for research and development of machine intelligence. The acid test would be to negotiate peace between the Israelis and the Palestinians.

Electronic Soapbox

As recently as the early 1990s, access to the public through the mass media was controlled by governments and large corporations. But although the cost of running governments and corporations keeps escalating, the plummeting cost of information technology is leveling the playing field and making it easier for common-interest groups and even individuals to have their say, too. People can make themselves heard without spending the huge sums traditionally laid out for advertising and political campaigns. Many transnational groups have organized and become powerful political influences through their activism on the World Wide Web.

By exposing political, environmental, and social abuses, these nongovernmental organizations are influencing policy on supranational issues, such as globalization, environmental pollution, climate change, and human rights.[3] The economic calculus of this leveling is changing the entire political landscape. Politicians, believing that they must spend fortunes on access to traditional mass media, are reduced to selling political favors, whereas citizens and activist groups can publicly support or challenge political rhetoric at practically no cost.

We have also seen a similar leveling effect in business. On-line forums provide consumer ratings and valuable feedback about commercial products and services. As this practice becomes more widespread, consumers may gain the upper hand over corporate advertising, misrepresentation, dangerous and unreliable products, and fraud. E-commerce allows small, upstart companies to compete effectively with big, established ones—the market shake-out of 2001–2002 notwithstanding. Will cheap information technology also allow small and poor *countries* to effectively compete in the world market with big ones, upsetting the traditional international calculus of power?

There is, of course, a dark side to a world in which information flows freely and unconstrained. What about information that is genuinely dangerous to society if not handled with great care? Free information can also empower extremist individuals and groups with violent agendas and who can cause substantial damage to society. Individuals can now more readily organize support and find camaraderie among other like-minded people. Terrorists form global networks that multiply the power of individuals and disseminate information about building and using a great variety of weapons. The information age opens the door to cyber-crime, cyber-terrorism, and cyber-warfare, against which we will need to develop elaborate defenses. Controlling access to dangerous knowledge (and deciding which knowledge is dangerous), while preserving the rights to free expression, will be one of the great challenges of future democracies.

No More Secrets

If you look around the world today, you see information in layers, sort of like an onion. There is the outermost layer, the sugar-coated, "official" version of reality that corporations, media, politicians, and governments want you to see. I call it the *bullshit layer*. If you peel away the layers in *Sixty Minutes* fashion, you slowly find out what is really going on—the information that those in power don't want you to know and that usually comes out only after they have left power. What would life be like without the bullshit layer? Would we even be able to deal with the plain, unvarnished truth? Open-information technology may offer some help.

Once in a while, there's a story or a movie (fiction, of course) about some politician who is suddenly compelled to tell only the truth. We all think it's funny, but strangely appealing, because we secretly long for a world where it's safe to tell the truth without having to worry about political correctness. But when some candidate for public office unambiguously declares his or her position on an emotional issue like gun control or abortion, the electorate becomes instantly polarized. The onion must be peeled very carefully!

Technology now makes it possible to broadcast information on the Web with complete anonymity. When there is no possibility of being traced, and the fear of retribution disappears, whistle-blowers come out of the woodwork. Anonymity makes it so easy to "publish" information that "no more secrets" could become a reality. If just one disillusioned employee feels that some government or corporate abuse or misdeed should be made public, then it is a simple and safe matter to make it so.

Of course, when the secrets turn out to be your credit-card numbers or your medical records, it is clear that more secure ways to conduct Internet transactions will have to be devised. An international cyber-crime treaty, now in the works, would make it impossible for Internet users to hide their on-line identities and activities. This treaty, if adopted in its present form, would set up a titanic clash between personal privacy and those who seek to control cyber-crime.

It is, of course, natural for those who occupy existing seats of power to fear and obstruct any leveling of the information playing field. Autocracies panic at the thought of increasingly vocal demands for democracy and open communication among their citizens, not to mention with outsiders—and well they should. Look what happened to the Soviet Union as a result of Mikhail Gorbachev's *glasnost* policy! China's autocratic government still has its fingers in the information dike, but it seems unable to stem the flood of information to and from its citizens.

Even corporations and government agencies combat efforts to open up communication among employees, fearing open criticism of policies and loss of control. But open communication should ultimately dissolve government and corporate secrecy and give ordinary people more say in their government, their workplace, and the marketplace. But this opening-up will probably not occur "within the system," as many expect, because the so-called system has too many incentives to keep things closed. Corruption and abuses of power will most likely be fought by being exposed to the light of public scrutiny. Governments and corporations that cannot function in the open will fall by the wayside.

The Napster Effect

The technology of the Internet will almost certainly lead to the end of intellectual property rights as we know them. Projects like Napster, Ian Clarke's Free Network Project (Freenet), and Publius preview how easy it will be to freely distribute music, pictures, books, movies, and other intellectual property. Clarke's idea is that the decentralization of information storage, along with suitable encryption of users' identities, will allow any document or any kind of media to be posted on the Net and retrieved completely anonymously, without fear of censorship and without a trace of any individual who can be held legally responsible. Since no single node in such a distributed information system is essential to its operation, it is relatively invulnerable to attack. Legal challenges and encrypted security schemes will produce temporary setbacks, but free information seems destined to become nearly universal.[4]

The social, legal, moral, and ethical implications of open information systems are astounding: Because a large chunk of our economy is based on the concept of intellectual property rights, many people will oppose any scheme that denies artists and other content creators much of the compensation that they deserve for their work. You will hear more and more arguments that open information systems are immoral and socially and economically disruptive, and those who make their living this way will push legislation to outlaw these schemes. But because this technology is inevitable, our traditional concepts of intellectual property rights will have to be rethought to accommodate it. In the future, information may well cease to be regarded as property.

Electronic democracy could provide every citizen with sufficient information and the means to participate meaningfully in government—from local to national. But as with all technology, the system can be abused. The temptation to tamper with the machinery for the purpose of manipulating the social system in Big Brother fashion will be irresistible. The same machinery can be used to disseminate propaganda, to filter and distort information to the extent that no citizen will believe anything the government or a politician says. Or perhaps, as envisioned by Huxley in *Brave New World*, information can be managed so skillfully that people *will* believe whatever they hear.

The outcome we get will depend not only on how strongly we assert our rights to open government and to free, reliable information, but also on who or what watches the machines.

21

Rethinking the Covenant
Between Science and Society

Decisions about the uses of science seem to demand a kind of wisdom which, for some curious reason, scientists are denied.

B. F. SKINNER

The story so far: Left to its own devices, the AI community is likely to produce marketable intelligent artifacts faster than they can be wisely assimilated by our economies and our societies. Creating machines that we cannot completely control and then putting them in charge of critical aspects of our lives poses risks whose consequences we may not have the luxury of contemplating after the fact.

Our experiences with nuclear weapons remind us that such warnings inevitably go unheeded, and that new technologies always take on a life of their own, resisting all attempts to coax them back into their bottles. A sensible policy would be to think a lot more about consequences up front, rather than playing it by ear as we go along. This kind of sensibility, however, seems contrary to human nature.

If there are other civilizations in the universe, then each one must reach a critical point during its development, when its science and technology first offer its citizens the means for their own destruction. Is it just luck that determines whether a civilization is swept into this vortex and annihilates itself, or is there some deliberate strategy that safely guides some through these perilous waters?

If there is such a strategy, then surely it is a part of *the covenant between science and society*. A sloppy or poorly thought-out relationship seems ultimately doomed, whereas one that wisely integrates scientific and societal values offers more hope for survival. This chapter, therefore, explores some seemingly dysfunctional aspects of our covenant between science and society, in the hope of finding something we can actually do to lessen our anxiety about accelerating technological change—including, but not limited to, AI.

Is Science to Blame?

Some see science as a threat to our survival. Vaclav Havel, president of the Czech Republic, wrote that science "describes the different ways we might destroy ourselves, but it cannot offer us truly effective and practicable instructions on how to avert them." The Strangelovian science that served U.S. and Soviet Cold War policies (mutually assured destruction, etc.) still haunts the public's minds as an example of science gone berserk. Today, deadly pollution accidents, the impact of technology on the global environment, and genetic experiments gone awry are easy to blame on science—and seem to validate Havel's lament.

Those who mistrust science see ominous threats in the future. They suggest that it is time to begin strictly controlling research in certain fields that seem particularly risky: genetic engineering, nanotechnology, and intelligent machines. Antiscience and back-to-nature movements go even further and urge us to abandon the scientific paradigm altogether—to somehow put a stop to the quest for new knowledge and to focus instead on learning to get along with each other.

Are these rational reactions to perceived threats posed by new scientific knowledge and new technology? Does it really make sense to blame new scientific discoveries for the social ills that our abuses of those discoveries beget? Is it all out of control, or are there ways that scientists, engineers, and thinking citizens at large might gain some control over the changes that such technologies will bring to their lives? And might science itself offer us any sensible, self-regulating and self-directing strategies for sustaining our civilization?

Knowledge Is Good

This mantra of Western science has driven us to explore and understand more and more about how nature works and to develop world-altering technologies. But exploration always involves risk, and new knowledge exacts its price. Those who probe the frontiers of knowledge (Magellan, Marie Curie, the *Challenger* crew) frequently pay with their lives. Since the middle of the twentieth century, however, the risks have taken on new and ominous proportions. In what sense can the power to destroy ourselves as a species be good?

Havel's complaint brings to mind the 1984 movie *The Gods Must Be Crazy*. It tells what happens to a small tribe of Kalahari bush people when a passing pilot discards an empty Coke bottle in their midst. At first, the tribe invents many uses for this novel item, which they regard as a gift from the gods. But soon they begin to scrap over possession of the bottle and discover the unfamiliar emotions of possessiveness and greed. When members of the tribe start to get hurt in fights, they finally reject the gift and seek ways to return the evil thing to the gods. It may not be good anthropology, but it's a wonderful parable about the consequences of introducing a technology into a society that is morally and ethically unprepared to deal with it.

Science is just a tool we've discovered for learning about nature. It comes with no instruction book for using its discoveries wisely. The problem is that new knowledge becomes dangerous when we are unprepared for its consequences. So far, we've dealt with such risks by trial and error, because no one has learned how to predict the consequences of new knowledge. But now, the consequences of really big mistakes are so great that we can no longer risk learning by the trial-and-error method. The *knowledge-is-good* paradigm worked fine during our scientific and technological childhood. But now we are struggling through a dangerous technological adolescence. The problem that typifies this developmental phase is not too much knowledge, but too little self-control.

The Technological Imperative

Our ambivalent relationship with science seems driven by a kind of technological imperative—one that presses us onward, to push to the max every technology that we can possibly profit from. The result has been that certain segments of the world's population have access to bigger, better, and faster global transportation and communication networks; spectacular entertainment; and improved health, medicine, and overall quality of life. But as the power of technology changes our lives, only a few of us stop to ask if these are the changes we want. As Bill Joy, cofounder of Sun Microsystems, put it:

> Failing to understand the consequences of our inventions while we are in the rapture of discovery and innovation seems to be a common fault of scientists and technologists; we have long been driven by the overarching desire to know. That is the nature of science's quest, not stopping to notice that the progress to newer and more powerful technologies can take on a life of its own.[1]

One invention that took on a life of its own was the atomic bomb. After the defeat of Nazi Germany, some Manhattan Project scientists questioned the morality of continuing to work on the bomb. But scientific and bureaucratic momentum driven by a wartime mentality won out. One atomic bomb was tested, then two bombs were dropped on Japan, a decision whose wisdom is still debated today. For half a century, the planet was held in the grip of a Cold War ethic based on threats of mutual annihilation by nuclear weapons.

Another area in which technology seems to take on a life of its own is medical science. We can now prolong almost indefinitely the lives of the terminally ill. We can, at great expense, substitute machines that function as artificial hearts, kidneys, and lungs, when our natural counterparts fail. We often do these things because we can—sometimes forgetting that we are prolonging human suffering and denying our loved ones a dignified death.

When a technology starts to get us into trouble by raising such sticky moral and ethical dilemmas, the usual response is opposition to the technology itself, rather than a call for clear and careful thinking about *new social, moral, and ethical structures* to accommodate it. Must our moral and ethical codes lag behind our science and technology, always leaving us trapped in ethical dilemmas? Or is there a better way?

Does Science Have a Conscience?

By asserting that science and technology are morally neutral and that moral judgments lie outside science's domain, scientists sometimes seem to divorce themselves from the rest of the human race. The chasm dividing science from the humanities that C. P. Snow described in *The Two Cultures* is as large today as it was in 1959.[2] B. F. Skinner's words at the beginning of this chapter suggest that most scientists are ill equipped and ill inclined to engage in discussions of the human and environmental consequences of their craft. Is this really true?

What other outcome would you expect of an education system that trains most scientists and engineers as technical specialists? Because their curricula, for the most part, leave out any preparation

for their roles as citizens and human beings, many young scientists
still leave their academic nests believing such traditional statements
about science as these:

- Science pursues knowledge for its own sake.
- Scientific inquiry knows no bounds.
- Science is rational, objective, and neutral.
- Human emotions have no place in the study of nature.
- Science is separate from politics.
- Science can't be blamed for its misuse.

These beliefs don't exactly instill a sense of social responsibility. In
fact, the logical consequence of this ivory-tower mentality is a lais-
sez-faire science policy: *Leave me alone to discover how nature works.
What society does with my discoveries is its business.* Scientists often
say that their job is to dispassionately present the scientific facts to
policymakers (or at least to their employers), and that moral judgments lie outside science's domain. It's not their fault, for example, if
the Internet is used to disseminate terrorist agendas, or if rockets are
used to deliver nuclear weapons. They would say that blaming science for society's ills is like blaming a chain saw for the consequences
of using it to trim your fingernails! This logic is strangely compelling
to scientists and technologists who simply don't want to be bothered
with the social, moral, and ethical implications of their work.

Once in a while, concerned and socially aware scientists and engineers speak out as the conscience of science. After the Manhattan
Project completed its task in 1945, a group of scientists, led by J.
Robert Oppenheimer, organized to recommend ways to keep the
nuclear genie in its bottle. One recommendation was that all work
on nuclear weapons be controlled by a single international agency,
such as the United Nations. But political forces and national rival-
ries allowed the genie to escape, and we live with the consequences
to this day.

Many scientists today believe that they share responsibility for the threat to humanity of nuclear weapons and refuse to work in the field—or even on weapons of any kind. But even more scientists believe that it is immoral *not* to summon every available technology for the cause of "national security." They see more technology as the only answer to the problems that the abuse of technology has already spawned.

The environmental movement of the 1970s awakened a sense of responsibility for the health of the planet and led to the establishment of government agencies like the U.S. Environmental Protection Agency (EPA). In the 1980s, groups of scientists organized to challenge the wisdom of President Reagan's Strategic Defense Initiative (Star Wars), and Carl Sagan led a crusade to warn us about the threat of the so-called nuclear winter. And in the 1990s, the world's climate scientists spoke out in a series of scientific summits to warn of global warming caused by industrial contributions to the earth's greenhouse gases.

At the birth of the twenty-first century, Bill Joy, speaking out as another conscience of science, took a thoughtful look at current trends in technology. He concluded that certain fields of inquiry will soon pose imminent danger to society—specifically, genetic engineering, nanotechnology, and robotics, or GNR for short. He recommended that we restrain the development of GNR until we figure out how to deal with the possible consequences, such as uncontrolled self-replication and competition with the human species.

The many critics of Joy's controversial article (which are easily found searching the Web) take issue not so much with his warnings as with his solution, which would turn the scientific paradigm of free inquiry on its head. They mainly stake out and defend the traditional scientific position that any restraint of free scientific inquiry is not only impossible, but not in society's interest. First, they say that any attempt to restrict a particular line of research will simply drive it underground: If something *can* be done, it *will* be done. And because the details of technological fallout are impossible to predict, indiscriminate restriction of some line of research would cut off beneficial as well as harmful outcomes. The use of lasers, for example, both as weapons and as medical tools is an acceptable trade-off; the risks (and financial rewards) of a "dark side" are balanced by the healing benefits. Joy's critics insist that history shows that science and technology flourish and are always most prolific and fruitful in a free setting. Maybe so, but these rebuttals merely defend the scientific status quo, and none really address Joy's warnings.

So, although science occasionally seems to have a conscience about the social ills that flow (or might flow) from new discoveries, the fraction of the scientific community that actually speaks out about the societal consequences of scientific work remains a small minority. Scientists are much more likely to speak out about funding cuts or attempts to regulate the scope of their craft.

Even the public and private institutions that employ and support scientists encourage scientific isolationism. Often, the conditions of their employment or grants actually forbid scientists to cross over and speak on the political or social implications of their work. Government and corporate scientists, for example, are usually forbidden to speak openly about the public policy implications of their research, unless those views are in harmony with the "party line."

Sometimes, the conscience of science even seems to hinder responsible scientific inquiry. Simply raising legitimate public safety issues surrounding a particular technology can result in public hysteria, even when there is no solid evidence to go on. Scientists rightly fear the popular media, which commonly sensationalize technological risks beyond reason. Because the public has trouble understanding scientific data, it is more receptive to emotional arguments. Most people have only a visceral reaction to the biological harm caused by electromagnetic fields and the cancer risks of exposure to certain industrial chemicals. Because of overexposure in the media, words like *radiation* and *chemical* carry their own emotional baggage.

The fierce debate over the risks and rewards of human cloning is another struggle between moral conscience, economic rewards, and scientific egos. The specter of designer babies and grisly accidents has led most governments (and the United Nations) to ban human cloning, pending careful study of the scientific, moral, and ethical issues. But you can be sure that a historic race to be the first is already under way in private, underground laboratories. The financial rewards are too great to be upstaged by moral and ethical issues, which will no doubt be cleaned up after the fact. How we handle human cloning will give us a preview of our chances of regulating even larger issues raised by intelligent machines.

Technology Assessment

Are the course of scientific advances and the directions of their technological fallout truly impossible to predict and control? Do we really have to accept the risks of a dark side, along with the beneficial side, of any given technology? If so, then are we just along for the ride, with little or no control over the directions that science and technology take us? The only responsible answers are no, no, and no.

In the early 1970s, there was a lot of interest in a new science called *technology assessment*. The idea was that, with sufficient care and imagination, one could forecast the short- and long-term societal and environmental impacts of new knowledge and new technologies. So equipped, policymakers were supposed to make informed judgments about where to invest public funds. The main practical fallout of this thinking in the United States was the formation of the EPA and numerous think tanks and the passage of laws requiring industry to assess the environmental impact of new commercial ventures.

But although millions of words were written about technology assessment, and even journals established to promote it, the tools for useful and objective technology forecasts simply did not exist at the time. (Nobody could have predicted the phenomenal growth and impact of the World Wide Web, for example.) Furthermore, some of the technological assessments that were actually performed required clearer thinking about societal goals and presented politically unpalatable choices. The U.S. Congress, apparently not inclined to deal with these choices, voted to close its Office of Technology Assessment in 1995. Consequently, the more ambitious aim of objective and accurate technological forecasting in the context of specific societal goals remains largely unrealized to this day.

What are the prospects for much better technology assessment tools in the future? And even if substantially better tools *are* developed, will we be politically and morally ready to make effective use of them? People are notoriously bad at forecasting the consequences of new knowledge. Respected "experts" asserted that a flying machine was impossible, that the sound barrier could never be broken, and that computers would never see widespread use. During their invention and early development, we could never have foreseen or designed the paths that semiconductors or lasers or rockets have taken. Today, it might be helpful to preview the societal and environmental impacts of, say, widespread telecommuting, or setting gasoline prices at five dollars a gallon, or charging five dollars per pound of household trash.

Might intelligent machines do a better job of simulating a realistic techno-socio-economic environment? After all, is exploring the innumerable branches on the tree of a developing technology very different, at its core, from playing chess? A rudimentary technology assessment tool already exists in Will Wright's popular PC game, Sim City. It lets you create and manage a simulated city and its services from scratch, balancing various economic, social, industrial, environmental, and quality-of-life factors. You can see the consequences of many choices: raising or lowering taxes; adding a factory, casino, or military base to your city; and dealing with various natural and human-caused disasters.

It is not difficult to imagine a much more sophisticated big brother of Sim City—a simulator that would incorporate the most advanced social and economic models, as well as models of the planetary environment. It would map out in time the far-flung social, economic, and environmental repercussions of any given combination of decisions. For example, one could examine the consequences of different energy policies—the effects of the balance of fuels we use, fossil-fuel exploration, exploiting alternative energy technologies, pricing, taxation, and conservation.

Experimenting with simulated societies could help us preview the unfolding future—not just the societal impact of particular technologies, but also the effects of changing moral values on those impacts. In other words, which ground rules for exploiting technology would produce the most beneficial outcomes for the simulated society? In this way, alternative moral and ethical codes that might offset our self-destructive tendencies could be devised and tested.

Developing such hugely complex forecasting programs would present formidable problems. First, such programs could well be quite unstable with respect to small changes in the details and assumptions of the models, that is, the rules of the game. The programs would also have to deal with the influence of unpredictable events. (Who could have predicted the impact of the Three Mile Island accident on attitudes toward nuclear power in the United States?) We should therefore not expect to be able to predict outcomes of a given technology in any great detail. But given the will to do so, we could design intelligent simulators that would give us a clearer view of alternative futures, if only in broad strokes. The cost of such a long-term undertaking would be huge, but probably only a fraction of current costs of preparing for war.

Science for Sale

"Western cultures, beginning with the Greeks, have been driven by an unbounded desire to know and explore. This desire, this insatiable curiosity, lies at the root of all scientific inquiry." Is this flowery preface to a research laboratory's brochure really true, or is it just a myth created for public consumption? What drives science and scientists? On a personal level, why do they do what they do? During my career, I have heard many different answers to these questions. Some careers are driven by a genuine burning curiosity about nature and a desire to make life better for us all. Others are more pragmatic and down-to-earth, reflecting the precedence of making a living over ideology. The most disturbing but pervasive reasons boil down to *It's a job,* and *That's what my boss told me to do,* as though scientists and engineers regarded themselves as parts of some giant machine. Some careers pass in turn through all these stages under the pressures of institutionalized science.

As much as we like to believe the popular image of science driven by the excitement of discovery and new knowledge and the challenge of serving humankind, a peek behind the scenes reveals something else. Individual scientists and scientific institutions are also driven by more myopic political and economic forces. Scientific priorities are set by governments and corporations with their own distinctly nonscientific agendas and ideologies. Consequently, scientists are often in the business of selling quick answers to politicians unwilling to risk meaningful but possibly painful social action. Influencing the directions of science and technology thus means taking part in the political process of setting public and corporate research priorities and budgets.

A huge chunk of public resources that we all have something to say about pays for military science and technology. Carl Sagan estimated that fully half of the scientists in the post–Cold War world are still employed at least part-time by the military. Many question whether this is a healthy allocation of our planet's science and engineering talent. Crime, economic polarization, environmental decay, a culture permeated by violence, and the failure of our education and penal systems would seem to pose far greater internal threats to national security than any barbarians at the gate.

The Cold War and the space race spawned a fifty-year boom in science and technology and created a substantial R&D community addicted not only to defense funds, but to military-style solutions to society's problems. "Threats to national security," manufactured to perpetuate Cold War largess, urge us to make a "war on drugs" and to arm ourselves against "rogue nations." President Dwight D. Eisenhower recognized that enemies are often conjured up by and for the welfare of what he called the military-industrial complex. Perhaps, more than forty years later, it is time to ask whether nations still require a well-defined enemy to justify a strong science program.

Apart from the military, who decides what directions science and technology take? Nonmilitary science funded by governments and corporations seems to be driven largely by perceived market demand. Is allowing science to be driven by market forces such a bad idea, particularly if we believe Skinner's words at the beginning of this chapter? Some argue that gearing science and technology to the marketplace automatically optimizes their long-term benefits to society. Is this really true or just free-market propaganda?

Allowing science and technology to be totally market driven gives the economic growth of the industrialized nations priority over society's most urgent problems. Instead of benefiting humanity, the system rewards those who provide products and services to those relatively well-off segments of society who can afford them. Survival in the marketplace is quite a different thing from assuring the survival and well-being of society as a whole. The long-term results of solely market-driven R&D are to promote overconsumption by the wealthy nations and to further polarize society into technological haves and have-nots. This polarization already threatens the security of the industrialized nations, whose complex technological infrastructure makes them vulnerable to unsophisticated attacks.

A concrete example of a polarizing effect of technology is provided by the pharmaceutical industry, whose R&D seems driven more by perceived profit than by humanitarian concerns or social need. Malaria, a rampant epidemic of developing countries, could be readily eradicated if there were the same kind of payoff for the drug industry as for Prozac or Viagra.

Many worthwhile technologies have the potential to generate tremendous social benefit, yet they remain glaringly underfunded because they lack obvious potential for short-term profit, because of public apathy, and because they threaten entrenched power structures. Alternative energy sources and clean, efficient transportation are oft-cited examples.

Another example of an underexploited technology with huge societal implications is so-called telepresence. Imagine replacing up to 95 percent of our personal transportation with a telecommunication system that would let people interact at a distance with full sensory experience indistinguishable from being there in person.[3] Benefits would include vast reductions in the energy consumption, pollution, urban congestion, and personal stress associated with commuting to work. Intelligent machines could deliver goods and dispose of waste. Education, recreation, and government could be transformed as well. And because moral behavior is prompted in part by empathy between individuals, telepresence could also help remove cultural and distance barriers that fuel conflicts. Of course, huge commercial empires can be expected to oppose such a massive reorientation of the vast transportation industry and its supporting infrastructure.

Socially Aware Science

If not the marketplace, if not the defense establishment, and if not the scientists themselves, then what forces might steer the research of scientists and the inventiveness of technologists in directions more likely to benefit humanity as a whole? How can we inject enough social wisdom and responsibility into a process that seems corrupted by political influence and for which profitability provides such strong incentives?

Such questions have to do with the interaction of science and public policy, a vast subject of many other treatises. Former EPA administrator William Ruckelshaus said that "taking control of the future means tightening the connection between science and public policy."[4] He creates the metaphor of a canoeist shooting the rapids: You need to know where the rocks are in time to steer around them. The knowledge is the science, and the steering is the policy.

What chance do you and I have of getting our hands on the paddles that steer R&D priorities? If we believe that public policy should truly have something to do with the public, then each of us can assume some individual responsibility for the science that we do, or at least some little part of the science we support as taxpayers. If science seems to be taking us somewhere that we don't want to go, then we can learn enough to help redirect it. Policymakers may sometimes seem inaccessible to the person on the street, but most are sensitive to consistent trends in public opinion. Becoming better informed *and* influencing policy are *both* facilitated by the information age. That's what the New Democracy and the redistribution of power are all about. Get involved. Ask questions. Question the answers. Demand an accounting. If scientists can't explain to you what they're doing and why, and what it costs, in terms you can understand, then they probably don't *know* what they're doing and certainly shouldn't have their hands in the taxpayers' pockets!

During World War II and the Cold War that followed it, the public and policymakers scarcely had to be sold on a strong science and technology program as a response to the Axis threat, or, even better, an Evil Empire. Criticizing weapons research was simply unpatriotic, and even if you did, security barriers kept you from asking too many questions. But the end of the Cold War and the advent of the information age place added responsibility on all of us to be informed about more complex science-based issues that affect us. Sagan wrote this about the public's responsibility for science policy:

> The technological perils that science serves up, its implicit challenge to received wisdom, and its perceived difficulty, are all reasons for some people to mistrust and avoid it . . . but the consequences of scientific illiteracy are far more dangerous in our time than at any time that has come before. . . . How can we affect national policy—or even make intelligent decisions in our own lives—if we don't grasp the underlying issues?[5]

Cleaning House

Fixing the dysfunctional relationship between science and society is not, as many scientists suggest, solely a matter of better public relations about scientific and technological issues. To ensure that publicly supported R&D is directed toward socially important problems, we also need to raise fundamental questions about how the R&D enterprise itself is presently funded and managed—and who stands to profit from it. According to science-policy analyst Daniel Sarewitz, several popularly held beliefs—including the following—permit science and technology to become less responsive to societal needs and should be reexamined:[6]

- More science and more technology will lead to more public good.
- Most scientific discoveries are made by accident, so if you just let scientists go where their curiosity leads them, you're bound to come up with some socially relevant results.
- Peer review, reproducibility of results, and other internal controls on the quality of scientific research address the main ethical responsibilities of the research system.

Scientists have been taught that these values underpin the integrity of their craft, but a closer look at today's institutionalized science suggests that these values may be more self-serving and less in the public interest.

Left alone in the post–Cold War setting, Darwinian selection seems to be producing not better science, but better salesmanship. The practice of institutionalized science tends to be less about acquiring knowledge in the service of humankind, and more about the survival, and even the enrichment, of its practitioners. Ethical concerns have shifted from the usual academic wrongdoing, such as falsification of data and plagiarism, to the misuse of sponsors' funds and even deceiving the public.

Consider this case in point. Pharmaceutical companies frequently sponsor research to test the effectiveness of their own products, but they often mask this sponsorship in the publication of the results. In published studies of the effectiveness of antidepressant drugs versus psychotherapy, the papers that claim better results for drugs are consistently authored by researchers paid by the drug companies, yet this sponsorship is often not mentioned in the papers and is thus kept from the public.[7] Imagine the pressures on researchers to emphasize favorable outcomes and tone down negative results, such as dangerous side effects, when their continued livelihood depends on pleasing their employer or sponsor.

What responsibility does the scientific community itself bear for looking critically at the entire business of science and coming up with a plan for real reform? Who should be responsible for assuring that the science enterprise reclaims its respectability and credibility?

Responsibility is shared by scientific leaders, scientific institutions, and individual scientists. Instead of calling for increased research budgets, *scientific leaders* might gain more public support by advocating housecleaning measures that would make national science programs more focused, efficient, and attuned to post-

Cold War realities. Our *scientific institutions* could open themselves to critical inquiry that examines waste and the scientific integrity of their programs. *Individually,* scientists can learn to recognize bogus science and ethical lapses when they see them and to question, even at the risk of their own livelihood, programs and institutions that waste resources or discourage (for example, by excessive secrecy) skeptical inquiry.

Here, then, are some specific ways the science community could regulate itself and assure its relevance to the society that supports it:

- Critically examine the connections between scientific research and the goals of society itself. This, of course, assumes that we can agree on, and have clear pictures of, those goals.
- Integrate research in the physical sciences with the desired economic and behavioral outcomes at the earliest stages of R&D projects.
- Replace secrecy and scientific arrogance with public dialogues about what types of scientific knowledge should be pursued. Open up science and technology policy to public input.
- Replace the mentality of undirected economic growth with the idea of sustainable development. This means tuning our science and technology to balance the developmental needs of the present with those of future generations.
- Prepare scientists and technologists more carefully for the social, economic, or political problems their craft is expected to address. We need more eloquent advocates for science and more scientists who are also educated in policy, law, economics, ethics, and communication.

In the future, science will not be able to hide behind its traditional curtain of mystery and elitism. It will have to justify its cost, not with doomsday scenarios, but by articulating clear visions of science in the public interest. And the public, in turn, can become full partners in shaping science policy by holding scientists accountable to broader societal goals. Science policy is about the choices we make, as individuals and as a society, about the kind of future we want. If we choose wisely, science and technology will serve us all well; if we do not, we will become their slaves.

Human Values and the Limits of Science

Havel's observation that science offers us more and better ways to destroy ourselves, but that it cannot teach us how to prevent their use, seems to describe a dysfunctional relationship between science and society. But is Havel's challenge a statement about the limits of the scientific paradigm itself, or just an observation that science comes with no instruction book to teach us how to use its results wisely?

The emergence of intelligent machines will surely present us with many novel ways to destroy ourselves. To prevent this from happening, we will have to think very carefully about what kinds of intelligent machines we want, which values will form their top-level instructions, and how to retain control. These decisions will require a critical and dispassionate (that is, scientific) examination of how well our own human values have served us. To do so, I believe that a *science of human values* is not only possible but imperative. By admitting the study of human values into the realm of legitimate scientific inquiry, we could bring all the critical investigative tools of science to bear on how our values shape our technological choices—for better or worse. *The goals of a science of human values would be to apply our critical scientific methods to understanding the bases of human ethics and morality, to rethink societal values and beliefs that seem dysfunctional in today's (and tomorrow's) world, and to design new values based on reason.*

Which human values bear closer scrutiny? Here are a few that affect how wisely we use technology:

- Our unquestioned beliefs about unlimited and undirected economic growth through the consumption of natural resources
- Our tendency to compete (often violently) rather than cooperate
- Patriotism, nationalism, and a tendency to follow leaders blindly
- Our inflated concepts of freedom and human dignity
- Our ideas of personal responsibility, crime, and punishment
- Certain self-serving myths promoted by the science and technology communities
- Our belief in the uniqueness of human intelligence

It has long been supposed that there are certain "ultimate questions" that lie beyond the reach of science—questions such as *Where did everything come from? Why is there matter rather than nothing? What is the meaning of life?* and *What happens after we die?* Many would add *How does the mind work?* and *What shapes human values?* Why are such questions so widely considered out of bounds for scientific inquiry? A likely reason is that the classic scientific method—forming falsifiable hypotheses and testing them by experiment—doesn't seem to work on them. Most people would have trouble forming testable hypotheses about the meaning of life, for example. But many such questions are just ill formed and can be reframed in ways that are testable and yield rational answers. Once upon a time, the list of such questions was much longer: *What is fire? What are the stars? Why does the earth shake? How does disease spread?* But the list shrinks as our curiosity leads us to rational answers for how nature works, and as mystical explanations give way to useful models. If we allow that science is just organized curiosity, then only our fears and superstitions shape its limits.

There are many cultural, political, and emotional reasons to fear a true science of human behavior—not to mention behavioral engineering! The potential for abuse is legendary. There is no shortage of critics to denounce it, because they say it can never account for this or that aspect of human behavior. It is fashionable in New Age circles to turn a deaf ear to anything "scientific" as cold and unfeeling—utterly unsuited to solving the human moral and ethical dilemmas that science itself seems to create. Even most scientists (and the institutions that support them) do not regard human values as a legitimate subject of scientific study. Richard Feynman, Nobel laureate and scientific icon of the twentieth century, asserted in his essay "The Value of Science," that "the question of the value of science is not a scientific subject."[8] (If not, then should the value of science be addressed in a nonscientific way?) And finally, our most powerful social institutions will fear the loss of control. How, then, shall we overcome the massive resistance to a scientific examination of human values?

The best near-term answer might be to avoid rather than overcome the political resistance by organizing privately funded research efforts. SETI (Search for Extra-Terrestrial Intelligence) is such an effort, and so were the construction and nonstop, round-the-world flight of Burt Rutan's *Voyager*. The staggering amount of private wealth in our society has the power to shape the future of humanity, yet much of this potential is squandered on self-indulgence.

Designing Our Future

Perhaps it is because Western science is supposed to be morally neutral that it so often fails to ask the question *What makes people happy?* In our relentless pursuit of scientific knowledge and material wealth, what we often discover instead is an emptiness within. By creating intelligent machines, we are forcing a confrontation with that emptiness, with our sense of purpose and self-worth, and with cherished, even sacred values. The outcome will determine the next steps in human evolution. Those steps can be of our own design, or we can surrender our fate to our creations.

History tells us that top-down-style global controls on the development and marketing of machine intelligence and other dangerous technologies are unlikely to work. But is rethinking our most sacred moral and ethical values any more likely? Maybe not, but history does tell us that ground-up movements, like the world's great religions, have a way of catching on, simply through the compelling power of their ideas.

22

What About God?

Since the subtitle of this book is *Intelligent Machines and Human Values*, I would be remiss if I failed to discuss the impact of AI on the most widespread human value on earth—the belief in God(s). Just as the appearance of superintelligent extraterrestrials on earth would cause humans to rethink their insular beliefs and perspectives, the (much more likely) revelation that we have created other thinking, conscious entities would cause many to examine more closely the factual basis of certain creation stories and the moral codes that rest on them. We have seen in Chapters 11 and 12 the pervasiveness of the social, moral, ethical, and legal implications of machine intelligence. As AI displaces mystical explanations for how the mind works, some people will rethink deeply held notions of what it means to be human, not to mention the role that God and religion play in their lives. Will the emergence of intelligent machines force a showdown between science and theology?

The vast majority of the world's population follows religions that, on the surface, urge people to treat each other with dignity and respect. Religions do offer comfort and solace in the face of mysterious and frightening aspects of nature—particularly suffering and death. Evolutionary psychologists view religion as one of the tools that evolution uses to encourage altruism, which game theory has shown to promote the survival of society as a whole. It follows that religious institutions survive and thrive because, like civil law, they promote social order and a stable environment in which to raise offspring. You might therefore expect religion to provide a stabilizing force in a species that likes explanations for things, but whose individual interests often conflict.

Then what went wrong? Why is the history of religion so full of tragic conflict—persecution, holy wars, and unspeakable atrocities committed in the name of God? How do religious messages about love and altruism get twisted into prejudice, hate, violence, and religious fanaticism? Are religions merely cultural organisms competing with each other, tooth and claw, for survival? If so, will the strongest eventually oust the others, will they reach some kind of peaceful accommodation, or must we endure ever-escalating religious conflict in the future?

The terrorist attacks on the United States not only shattered our complacency, but also sensitized us to the kinds of pernicious religious values (i.e., not our own!) that drive such acts. Whether malignant distortions or natural consequences of religious teachings, these acts motivate some to ask: *What sort of God instructs one people to hate another people so much?* But some go even further and ask a shocking question: *Have religious values and religious thinking themselves, with their polarizing effects, finally outlived their usefulness and become dysfunctional in today's emerging global community?* Is it simply time to scrap them altogether, or is there a rational framework that would allow us to hold on to the stabilizing and comforting influences of religious beliefs while discarding the destructive and divisive ones?

Why We Believe

Could understanding the mythological underpinnings of religious dogma—and why the human mind seems to need them—better equip us to choose rationally which beliefs to accept and which ones to discard? Why do people need God?

Joseph Campbell's discussions with Bill Moyers in the PBS series *The Power of Myth* offered a compelling account of the mythological bases for the world's religions: Every society constructs creation stories. These myths differ widely from culture to culture, but they reveal a common thread throughout our species: Our brains seem to be wired with a burning curiosity about our own origins, our ultimate fate, and about how nature works. Primitive creation myths were crude stories passed orally from generation to generation. It was natural to personalize the wrath of nature as the wrath of powerful, angry gods. As each generation embellished and spread its stories, they grew into legends, then rules for appeasing various deities (theologies), and finally into today's huge religious institutions, with their elaborate moral codes. The power of myth, then, culminates in the force it exerts over the lives of almost everyone on earth—believers and nonbelievers alike.

Most religious decisions were made for us as children learning at our parents' or teachers' knees—at an impressionable age when our minds are receptive to anything we are told and do not question authority. The way such values are passed from generation to generation, and our general resistance to changing such core values, accounts for regional uniformity in religious beliefs and practices. The irony is that, even though one's particular religious beliefs depend more on heredity and geographical accident than on reason and choice, people are often prepared to die (or kill) to uphold and defend them.[1]

For both primitive and modern humans, some religious myths are comforting notions, particularly the ones about a pleasurable afterlife. These myths are designed to assuage the greatest fear of all, the fear of death. But as explanations for how nature actually works, they flunk miserably. Most religions portray God as the all-knowing, all-powerful being(s) who created and runs the universe. Yet this notion of God *explains nothing* about how nature works—that is, it has no predictive power whatsoever. As for the Big Questions, no religion offers any credible or verifiable evidence about where we come from or where we go when we die. Instead, religious beliefs merely add superfluous layers of mystery on top of those already posed by nature.

The Machinery of Nature

In parallel with these myths, primitive man must have acquired lots of practical predictive knowledge every day: When dark clouds approach, it is likely to rain, or if his spear strikes a buffalo in a certain spot, the animal will be killed. This kind of knowledge is learned by *experience*, that is, by trial and error. Such knowledge is, of course, the beginning of *science*. The classical Greeks were probably the first culture to formalize the scientific method of observing nature and testing ideas to see if they work. Today, as their legacy, we have so much practical knowledge about how nature works that it is institutionalized and codified, and no one person can possibly know it all. To be sure, there are still lots of things about nature that we don't know, like *Why is there matter rather than nothing? Why are the laws of nature the way they are, rather than some other way? What is time? How does the mind work?* But these are merely gaps in our present knowledge, not gaps in the scope of the scientific method.

Because of its predictive successes, science has largely displaced religious beliefs as a viable explanation for how nature works. But because cherished myths die slowly, to this day we find ourselves seeking some kind of reconciliation between science and theology.

Many believe that the two can be reconciled because they are concerned with separate planes of experience (the spiritual and the material). But there are clearly areas of overlap for which the two offer utterly incompatible accounts. Is the earth a giant platform supported by eight mighty elephants or a sphere of accumulated stellar debris? Does creationism or evolution correctly account for the diversity of life on earth? Is the mind a marvelous machine that follows entirely physical laws, or is it a separate, nonphysical reality?

It has been said that science itself is a religion—that we simply accept the word of scientists, who create myths about how the physical world works.[2] Many people (including some scientists) do, in fact, accept certain scientific "facts" without question. Scientists have even been known to cling blindly to a favorite theory. But the difference between science and religion is that scientific beliefs are always (in principle) open to skeptical inquiry and testing in the real world. Religions that claim a monopoly on the truth and are absolutely certain about what is right and wrong merely define those who do not accept that truth as enemies. Any science (Christian Science and Scientology come to mind) that is not testable and open to question and criticism is not a science.

History records many examples of clashes between scientific observations and prevailing political or religious agendas, in which science was the loser. Galileo's and Copernicus's observations of planetary motions were declared heresy because they did not conform to religious dogma. T. D. Lysenko derailed advances in genetics for decades in Russia, because his sloppy science happened to support Marxist ideology. Today, energy producers (and consumers) downplay the scientific evidence that humans are causing global warming, because acknowledging it would require them to change the way they do business. Sometimes it is hard to distinguish between science in the service of society and science dictated by ideology and expediency (Chapter 21).

Mostly Harmless?

It is sensible to ask what harm there is in allowing religious beliefs to comfort us, even though they may not be scientifically verifiable or make strict logical sense. The harm lies in the grip that religious thinking holds on people's minds.

First, religious institutions today seem designed more to manipulate, control, and divide peoples than to inspire and unite them. While preaching love and harmony within their own ranks, they exclude (and even condemn) outsiders and incite conflicts with nonbelievers. Hyperpatriotism is a currently fashionable form of groupthink that, like religious fanaticism, substitutes righteousness for reason. To me, it seems quite indistinguishable from the forces that drive terrorism.

Second, as biologist Richard Dawkins put it, "I am against religion because it teaches us to be satisfied with not understanding the world." Faith-based ignorance about how the world works leaves us ill equipped to deal with the moral and ethical dilemmas (like stem-cell research and climate change) that technology continues to pose.

Third, religious dogma discourages independent thought and makes people easy prey for earthly autocrats. The short-lived Taliban regime in Afghanistan used Islamic orthodoxy to control the populace, while its leaders lived in luxury.

And fourth, religious doctrine can be used to justify any behavior whatsoever—torture, murder, even genocide. Once the distinction between good and evil—between "us" and "them"—becomes stark and absolute, then conquest and war are inevitable. *For God and country* or *In the name of Allah* become convenient battlefield slogans.

Wheat and Chaff

If religion fails to explain anything useful about nature, if it seems so much like mind control, and if it has so many disruptive social effects, then what of value is left? In what sense could religions possibly be good for people or societies today?

One could argue that religions inspire lots of altruistic activities, such as relief and aid organizations, that are good for society, but whose accomplishments go unheralded, eclipsed by headlines about religious fanaticism. Could people exhibit such altruism without religious inspiration? Although it is impossible to do this experiment, there are lots of rational sociological reasons for altruistic behavior that have nothing to do with beliefs in a supreme being. Religion is not only a nonessential component of these benefits; it has been known to inflict great harm. Many primitive societies would attest to the damage done by the religious strings that are often attached to well-meaning but overzealous aid.

Another argument for the positive effects of religion would claim that moral behavior is inspired by religious myths and rituals and that crimes are often inhibited by threats of religious retribution. We know that myths, fables, and legends have always inspired, and will no doubt continue to inspire, people to excel, to dream of artistic (and even scientific) achievements, to call upon their inner strengths and resources, and to build constructive relationships with nature and with other people. From *Little Red Riding Hood* to *Hamlet*, stories and legends often convey useful moral lessons. But none of these benefits depends on the stories' being literally true! Confusing these lessons with the truth is like ardent soap-opera or *Star Trek* fans who can't tell the difference between TV characters and real life.

In the same way, religious stories may be entertaining, inspiring, uplifting, and morally instructive, but there is no need to believe that any of them is literally true. In fact, much of religious doctrine is demonstrably false. Creationism, for example, bases its case on wisdom revealed by the Bible—a logical fallacy known as *argument from authority*, and which characterizes most religious teachings. Arguments from authority are not scientifically acceptable forms of reasoning, because they do not make predictions that can be objectively tested. Such arguments, in fact, discourage such testing as irrelevant.

So, although religious stories may be entertaining, inspiring, up-lifting, and morally instructive, their truth is not essential to their benefits. If we decided instead to regard these stories in the same way that we regard other literary fiction, then religious stories could continue to stir the soul and bind people together, inspiring them in times of need or distress. Then we could enjoy their beauty, ponder their moral lessons, and derive comfort from their insights into human nature.

Is it possible, in practice, for ordinary people to retain cherished myths, fables, and legends as inspirational art, while discarding the authoritarian control and prejudices of institutionalized religion? Such a rational approach to separating the two ignores the emo-tional attachment most people have to religion, its rituals, and sym-bols. The intelligentsia find it easy to say *This makes no sense!* and to substitute reason for faith, but most of the people in the world are so poor, sick, and uneducated that they are unable (and unmoti-vated) to reason clearly or strongly enough to overcome childhood indoctrination. For the least fortunate, religious beliefs are all they have to make their miserable existence tolerable. In short, the uned-ucated masses look to God for comfort, to keep them out of trouble, and even to tell them what to think.

Any useful change in religion's polarizing effects would have to begin with religions' backing off from their monopolies on the truth and acknowledging that religion is more like art: Its meaning and interpretation lie in the mind of the beholder. Unfortunately, the radical institutional and doctrinal changes required to kindle interfaith harmony would also undermine the power of religious authorities, who would naturally oppose them. The ecumenical pronouncements made so far by the world's religious leaders are pitifully inadequate. Perhaps, then, the responsibility for real change rests with parents and teachers. Instead of indoctrinating children, would it be that much harder to teach them to reason, observe, and think for themselves? Then they would be equipped to choose rationally which beliefs to accept and which to discard.

What About Prayer?

Western religions encourage the faithful to call upon God to grant their wishes—a kind of grown-up version of asking Santa Claus for toys as children. Does prayer work? We can't possibly know, because God never provides feedback. If we get what we ask for, how do we know that we wouldn't have received it anyway? If we do not get what we ask for, then it can always be said that God (or Santa Claus) *did* answer our prayers—but just said no.

Is prayer a rational or healthy practice? It seems to help believers cope with situations completely beyond their control by providing the illusion that they can regain some degree of control by bargaining with God. Is this self-deception a good idea? People placed in stressful situations whose outcome is totally beyond their control may experience anxiety, depression, hysteria, or physical illness. Asking God to spare the life of a dying loved one or to bring our children back safely from war may ease these harmful responses. Prayer allows us to believe that we are doing something effective, such as transferring responsibility for life and death out of earthly hands and into more responsible and compassionate hands. Even a simple "God help us!" spoken during a crisis can relieve tension. Prayer's feel-good benefits, then, seem akin to the calming effects of meditation.

But relying solely on prayer can produce disastrous results if doing so lets us avoid personal responsibilities, such as seeking appropriate medical care. Praying to win the lottery (a situation truly out of our control) may be a harmless exercise, but if trusting God to provide for our families allows us to avoid looking for a job, or if we ask God to bless our aircraft instead of servicing them, then we are just asking for trouble.

East and West

If the issue underlying prayer is control over our lives, then Eastern and Western religions take opposite views. Asian religions like Hinduism and Buddhism embrace a kind of fatalism—that all is in God's hands and beyond human control. Their prayers are more like submission to their destiny than pleas for help. What could be the survival value of such a belief? Shouldn't its practitioners soon die off from untreated starvation and disease? In fact, many do, but uncontrolled reproduction assures their replacement. Their numbers (but not their misery) are limited only by finite resources—by God, if you will.

We can look at fatalism as the Eastern response to the dilemma of free will (Chapter 7), namely, *Are we in control or not?* The faith of the West, by contrast, rests on a belief in human control, perhaps at a cost of self-indulgence, overconsumption, and a certain disharmony with nature. Its practitioners' numbers are limited by more subtle forces that affect the planet as a whole.

The fatalism of the East tends to undervalue individual human life, whereas Western illusions about control tend to overvalue it. As a result, Eastern cultures would probably relinquish control to superintelligent machines more readily than the West would. The clashes that arise at the interface between East and West—in the Middle East, for example—are about more than regional theology. They are nothing less than conflicts between opposite worldviews. Eastern problems are not amenable to Western solutions, and vice versa.

Faith or Reason?

Is understanding the world the prime driver of social progress? Is discovery the only measure of human advancement? Many say that *faith* is a more powerful force than reason, that *spiritual values*, not science and technology, show the path to humankind's ultimate happiness. Their faith tells them, for example, that human life is the sacred centerpiece of creation. Consequently, no matter how "smart" our machines get, they can never have a soul and never usurp our special relationship with God.

Cultural traditions have elevated faith to a certain level of legitimacy. After all, it is neat, simple, comforting, and difficult to argue with. Yet there is a fundamental problem with believing in things for which there is no evidence: *How shall we decide what to believe and what not to believe?* There is an infinite number of things we could possibly believe in: angels, ghosts, psychic powers, reincarnation, astrology, time travel, the healing power of crystals, and so forth. Believing in such things may bring us a kind of happiness, but it is the happiness of ignorance and self-delusion.

How shall we sort out the "truth" from illusion, myth, deception, and propaganda? If we say we shall believe what is written in some sacred book, such as the Bible or the Koran or *The Thoughts of Chairman Mao*, then how shall we decide which book to believe? And how shall we resolve contradictions within each book? Faith gives us no way to sort these things out. Instead, it provides what many people want—an excuse for not having to observe, think, weigh evidence, and decide for themselves. It substitutes arguments from authority for reason.

Gods for the Future

What truths might serve as more stable and universal foundations for the moral and ethical decisions facing our global society? Without religion, where shall we look for comfort and inspiration? What commandments will make us behave? And how can we possibly come to terms with our own mortality? The religious metaphors we choose reflect our individual needs for explanations, guidance, comfort, and motivation—or the lack of the same. They can move us with understanding and compassion for others, or they can fill our minds with a righteous need to control others.

Instead of regarding God as an old man with a white beard who lives in heaven, some people see God as "the force behind the universe," "the essence of life," or "the spirit that dwells within each of us," or as the Bible says, "God is love." Such metaphors give God more latitude and free the mind from constricting religious dogma. Many believe that God is not *out there* but *within us*—an expression of our inner yearning to grasp the wonder and mystery of life. For them, salvation lies not beyond the grave but in realizing one's potential and discovering a unity with nature and humanity in this life. Immortality lies in the hearts and minds of loved ones we leave behind. This God represents the inner resourcefulness that we call upon when we need it most. This idea of an internal God is stronger in Eastern religions, particularly in regions that are poor in worldly resources. The Sanskrit word *Namaste*, used as a salutation in northern India, Nepal, and Tibet, means roughly, "I bow to the divine in you."

In a future world of global intelligence and a global nervous system, the prevalent myth may become the oneness of the planet with its human tenants. Ideas like the Gaia hypothesis (that the earth and all its life act like a single 4-billion-year-old, self-regulating organism) may replace God as the inspirational metaphor that instills in us a sense of responsibility for each other and for the health of the planet as a whole.

God as Superintelligence

If scientists who seek to create AI are "playing God," then what if they succeed? Exposing the myth that humans are the only sentient, self-aware beings on the earth would undermine every human-centered concept of morality, man's place in nature, and his special relationship with God. How could such an upheaval of human affairs come about?

It seems inevitable that sometime in this century, Moore's Law (Chapter 4), combined with greater understanding of intelligence itself, will drive machine intelligence to levels beyond our own—and soon thereafter, beyond anything we can imagine. When intelligent machines begin designing superintelligent machines and their software, intelligence should grow exponentially. The result could be a runaway intelligence explosion comparable to the Cambrian explosion of life on earth, the biological equivalent of the Big Bang. Some call this coming intelligence explosion a *technological singularity*.[3] At this point, the available computational power in the world would, for any practical purpose, be infinite. Present levels of human intelligence would be insignificant by comparison. It would make as much sense to speak of people *using* computers as tools as it would to say that cows use humans as tools. The consequences for human dignity and our primacy on earth are so unpredictable that we cannot see beyond such a singularity.

Would this future need *us*? Even today, we are beginning to endow our governments and corporations with godlike power and authority. As they become more technologically enhanced and powerful (but not necessarily wiser), could they someday decide that they no longer need their citizens? If so, then human survival would likely depend on joining with intelligent machines to become a new life-form (Chapter 18). Vernor Vinge speaks of *intelligence amplification* (IA), a kind of human–computer symbiosis in which our brains are augmented by computers, and human minds are linked worldwide by broadband connections. Such technologies would not only permit virtual immortality (not necessarily a blessing), but would also blow away the distinction between individual human egos, and with them our ideas of personal rights and responsibilities. As a result, IA also sets out a minefield for just about every religious idea and moral concept that we hold dear.

The impact of an intelligence explosion on the human condition would be as far-reaching and unimaginable as if extraterrestrial beings with technology a million years ahead of ours appeared on earth. Any sufficiently advanced superintelligence, whether terrestrial or extraterrestrial, would seem incomprehensible, omniscient, and omnipotent to us—exactly the properties we ascribe to God. As physicist Freeman Dyson said, "God is what the mind becomes when it has passed beyond the scale of our comprehension."[4]

notes

....

Preface

1. D. Wooldridge, *The Machinery of the Brain* (New York: McGraw-Hill, 1963).
2. D. Wooldridge, *The Machinery of Life* (New York: McGraw-Hill, 1966); and D. Wooldridge, *Mechanical Man: The Physical Basis of Intelligent Life* (New York: McGraw-Hill, 1968).
3. H. Smith, *The Religions of Man* (New York: Harper and Row, 1958); B. F. Skinner, *Beyond Freedom and Dignity* (New York: Knopf, 1971); E. O. Wilson, *On Human Nature* (Cambridge: Harvard University Press, 1978); P. McCorduck, *Machines Who Think* (New York: W. H. Freeman and Co., 1979); D. Hofstadter, *Gödel, Escher, Bach: The Eternal Golden Braid* (New York: Basic Books, 1979); and R. Wright, *The Moral Animal* (New York: Pantheon, 1997).

Chapter 1: Artificial Intelligence—That's the Fake Kind, Right?

1. P. McCorduck, *Machines Who Think* (New York: W. H. Freeman, 1979).
2. J. Cohen, *Human Robots in Myth and Science* (London: Allen and Unwin, 1966).
3. J. Hogan, *Mind Matters: Exploring the World of Artificial Intelligence* (New York: Del Rey, 1998).
4. B. Joy, "Why the Future Doesn't Need Us," *Wired* 4 (April 2000): 238–262.

Chapter 2: What Makes Computers So Smart?

1. M. Minsky, *The Society of Mind* (New York: Simon and Schuster, 1985).

Chapter 3: What Do You Mean, Smarter than Us?

1. K. Warwick, *In the Mind of the Machine: Breakthrough in Artificial Intelligence* (London: Arrow Books, 1998).

2. A. Turing, "Computing Machinery and Intelligence," *Mind* 59, no. 236 (1950): 433–460. This classic article is widely reprinted on-line.

3. Dennett, D., "Can Machines Think?" essay no. 1 in *Brainchildren: Essays on Designing Minds* (Cambridge: MIT Press, 1998).

Chapter 4: Machines Who Think

1. P. McCorduck, *Machines Who Think* (New York: W. H. Freeman, 1979).

2. A. Turing, "Computing Machinery and Intelligence," *Mind* 59, no. 236 (1950): 433–460. This classic article is widely reprinted on-line.

3. D. Dennett, "Can Machines Think?" essay no. 1 in *Brainchildren: Essays on Designing Minds* (Cambridge: MIT Press, 1998).

4. H. Moravec, "When Will Computer Hardware Match the Human Brain?" *Journal of Transhumanism* 1 (March 1998). Available on-line.

5. P. Harmon and D. King, *Expert Systems* (New York: John Wiley and Sons, 1985).

6. J. Searle, "Minds, Brains, and Programs," in *The Mind's I*, ed. D. Hofstadter and D. Dennett (New York: Basic Books, 1981).

7. M. Sipper and J. Reggia, "Go Forth and Replicate," *Scientific American*, August 2001, 35–43.

8. S. Wolfram, *A New Kind of Science* (Champaign, Ill.: Wolfram Media, 2002).

Chapter 5: Let the Android Do It

1. I. Asimov, *I, Robot* (New York: Doubleday, 1950).

2. H. Moravec, "Rise of the Robots," *Scientific American*, December 1999, 124–135.

3. K. M. Ford and P. J. Hayes, "On Computational Wings: Rethinking the Goals of Artificial Intelligence," *Scientific American Presents: Machine Intelligence* 9 (winter 1998): 78–83.

4. H. Moravec, *Robot: Mere Machine to Transcendent Mind* (Oxford: Oxford University Press, 1999).

Chapter 6: What Is Intelligence?

1. K. M. Ford and P. J. Hayes, "On Computational Wings: Rethinking the Goals of Artificial Intelligence," *Scientific American Presents: Machine Intelligence* 9 (winter 1998): 78–83.

2. T. Kuhn, *The Structure of Scientific Revolutions* (Chicago: University of Chicago Press, 1962).

Chapter 7: What Is Consciousness?

1. M. Minsky, "Minds Are Simply What Brains Do." (on-line essay).
2. D. Hofstadter, *Gödel, Escher, Bach: An Eternal Golden Braid* (New York: Vintage Press, 1979).
3. V. Ramachandran and S. Blakeslee, *Phantoms in the Brain: Probing the Mysteries of the Human Mind* (New York: William Morrow, 1998).
4. R. Penrose, *The Emperor's New Mind* (Oxford: Oxford University Press, 1989); and R. Penrose, *Shadows of the Mind* (Oxford: Oxford University Press, 1994).
5. Hofstadter, *Gödel, Escher, Bach.*
6. B. F. Skinner, *Beyond Freedom and Dignity* (New York: Knopf, 1971).
7. M. Minsky, "Consciousness Is a Big Suitcase." Available on-line at www.edge.org.
8. R. Wright, *The Moral Animal* (New York: Pantheon, 1997), chapter 13.
9. E. Fredkin, "On the Soul." Draft manuscript, available on-line.

Chapter 8: Can Computers Have Emotions?

1. R. Picard, "Does HAL Cry Digital Tears? Emotions and Computers," chapter 13 of *HAL's Legacy* (Cambridge: MIT Press, 1997).
2. M. Minsky, "Consciousness Is a Big Suitcase." Available on-line at www.edge.org.
3. M. Minsky, *The Society of Mind* (New York: Simon and Schuster, 1986).
4. M. Minsky, *The Society of Mind.*
5. R. Picard, *Affective Computing* (Cambridge: MIT Press, 1997).

Chapter 9: Can Your PC Become Neurotic?

1. E. Dexler, *Engines of Creation* (Garden City, N.J.: Anchor Books, 1987).
2. B. Joy, "Why the Future Doesn't Need Us," *Wired* 8 (April 2000): 238–262.

Chapter 10: The Moral Mind

1. R. Wright, "Darwin Comes of Age," chapter 1 in *The Moral Animal* (New York: Pantheon, 1997), 27.
2. B. F. Skinner, *Beyond Freedom and Dignity* (New York: Knopf, 1971).

Chapter 11: Moral Problems with Intelligent Artifacts

1. M. Boden, *Artificial Intelligence and Natural Man* (Cambridge: MIT Press, 1987).
2. K. Ford, C. Glymour, P. J. Hayes, and K. Ford, eds., *Android Epistemology* (Cambridge: MIT Press, 1995).
3. D. Dennett, "When HAL Kills, Who's to Blame?" chapter 16 of *HAL's Legacy* (Cambridge: MIT Press, 1996).

Chapter 12: The Moral Machine

1. C. Darwin, *The Descent of Man, and Selection in Relation to Sex* (New York: Modern Library, 1871), chapter 12.
2. I. Asimov, "Runaround," *Astounding Science Fiction* (March 1942).
3. M. Minsky, "Will Robots Inherit the Earth?" *Scientific American,* October 1994, 108–113.
4. D. C. Dennett, "When HAL Kills, Who's to Blame?" chapter 16 of *HAL's Legacy* (Cambridge: MIT Press, 1997).

Chapter 13: Global Network to Global Mind

1. M. Brooks, "Global Brain," *New Scientist,* 24 June 2000, 22–27.
2. T. Berners-Lee, J. Hendler, and O. Lassila, "The Semantic Web," *Scientific American* 284 (May 2001): 34–43.

Chapter 14: Will Machines Take Over?

1. K. Warwick, *In the Mind of the Machine: Breakthrough in Artificial Intelligence* (London: Arrow Books, 1998); and J. Weizenbaum, *Computer Power and Human Reason* (New York: W. H. Freeman, 1976).

Chapter 16: Cultures in Collision

1. J. Diamond, *Guns, Germs, and Steel: The Fates of Human Societies* (New York: Norton, 1997).
2. C. P. Snow, *The Two Cultures* (Cambridge, UK: Cambridge University Press, 1993).

Chapter 17: Beyond Human Dignity

1. J. Weizenbaum, *Computer Power and Human Reason* (New York: W. H. Freeman, 1976).
2. B. F. Skinner, *Science and Human Behavior* (New York: Macmillan, 1953).

3. B. F. Skinner, *Beyond Freedom and Dignity* (New York: Knopf, 1971).

4. D. Wooldridge, *Mechanical Man: The Physical Basis of Intelligent Life* (New York: McGraw-Hill, 1968).

5. D. L. Nolte, "Children Learn What They Live," in *Children Learn What They Live: Parenting to Inspire Values,* by D. L. Nolte and R. Harris (New York: Workman Publishing, 1998).

Chapter 18: Extinction or Immortality?

1. R. Ornstein, *The Psychology of Consciousness* (New York: Penguin Books, 1986).

2. H. Moravec, *Mind Children: The Future of Robot and Human Intelligence* (Cambridge: Harvard University Press, 1990).

3. E. Regis, *Great Mambo Chicken and the Transhuman Condition* (New York: Addison-Wesley, 1990).

4. I. Crawford, "Where Are They?" *Scientific American,* July 2000, 38–43.

Chapter 19: The Enemy Within

1. C. Sagan, *The Demon-Haunted World: Science as a Candle in the Dark* (New York: Ballantine, 1996); and T. Schick, Jr., and L. Vaughn, *How to Think About Weird Things: Critical Thinking for a New Age* (New York: McGraw-Hill, 1995).

Chapter 20: Electronic Democracy

1. J. Alter, "The Couch Potato Vote: Soon, You'll Be Able to Vote from Home—But Should You?" *Newsweek,* 27 February 1995.

2. R. Fisher and W. Ury, *Getting to Yes: Negotiating Agreement Without Giving In* (New York: Penguin, 1981).

3. R. Wright, *Nonzero: The Logic of Human Destiny* (New York: Pantheon, 2000).

4. C. Mann, "Who Will Own Your Next Good Idea?" *Atlantic Monthly,* September 1998, 57–82.

Chapter 21: Rethinking the Covenant Between Science and Society

1. B. Joy, "Why the Future Doesn't Need Us," *Wired* 8 (April 2000): 238–262.

2. C. P. Snow, *The Two Cultures* (Cambridge, UK: Cambridge University Press, 1992).

3. J. Lanier, "Virtually There," *Scientific American,* April 2001, 66–77.

4. W. Ruckelshaus, "Toward a Sustainable World," *Scientific American*, September 1989, 166–174.

5. C. Sagan, *The Demon-Haunted World: Science as a Candle in the Dark* (New York: Random House, 1996).

6. D. Sarewitz, *Frontiers of Illusion: Science, Technology, and the Politics of Progress* (Philadelphia: Temple University Press, 1996).

7. S. Rampton and J. Stauber, *Trust Us, We're Experts!* (New York: Tarcher-Putnam, 2001).

8. R. Feynman, *The Pleasure of Finding Things Out: The Best Short Works* (New York: Penguin, 1999).

Chapter 22: What About God?

1. R. Dawkins, "The Emptiness of Theology." Essay available on the Web.

2. R. Dawkins, "Is Science a Religion?" Essay available on the Web.

3. V. Vinge, "The Technological Singularity," *Whole Earth Review,* winter 1993, 88–95.

4. F. Dyson, *Infinite in All Directions* (New York: Harper and Row, 1988).

about the author

Thomas M. Georges is a retired physicist living in Boulder, Colorado. From 1963 until 2000, he worked as a research scientist, first at the National Bureau of Standards, then at the Institute for Telecommunication Sciences, and finally with the National Oceanic and Atmospheric Administration. His research produced more than one hundred publications on environmental science and remote sensing, as well as a book about scientific and technical writing. His education includes a B.S. from Loyola-Marymount University, an M.S. from the University of California, Los Angeles, and a Ph.D. from the University of Colorado, all in electrical engineering.

index